MOBILE DATA MANAGEMENT
AND APPLICATIONS

MOBILE DATA MANAGEMENT
AND APPLICATIONS

edited by

Jin Jing
GTE Laboratories, USA

Anupam Joshi
University of Missouri, Columbia, USA

A Special Issue of
DISTRIBUTED AND PARALLEL DATABASES
Volume 7, No. 3 (1999)

KLUWER ACADEMIC PUBLISHERS
Boston / Dordrecht / London

Distributors for North, Central and South America:
Kluwer Academic Publishers
101 Philip Drive
Assinippi Park
Norwell, Massachusetts 02061 USA
Telephone (781) 871-6600
Fax (781) 871-6528
E-Mail <kluwer@wkap.com>

Distributors for all other countries:
Kluwer Academic Publishers Group
Distribution Centre
Post Office Box 322
3300 AH Dordrecht, THE NETHERLANDS
Telephone 31 78 6392 392
Fax 31 78 6546 474
E-Mail <services@wkap.nl>

 Electronic Services <http://www.wkap.nl>

Library of Congress Cataloging-in-Publication Data

Mobile data management and applications / edited by Jin Jing, Anupam
    Joshi.
       p.   cm.
    "A Special issue of Distributed and parallel databases, volume 7,
No. 3 (1999)."
    Includes bibliographical references.
    ISBN 0-7923-8596-9 (alk. paper)
    1. Mobile computing. 2. Database management. I. Jing, Jin.
II. Joshi, Anupam.
    QA76.59.M65   1999
    006.3—dc21   99-33653
                                                   CIP

Printed on acid-free paper.

Printed in the United States of America

DISTRIBUTED
AND
PARALLEL
DATABASES

Volume 7, No. 3, July 1999

Special Issue: Mobile Data Management and Applications
Guest Editors: Jin Jing and Anupam Joshi

Distributed and Parallel Databases is abstracted and/or indexed in *ISI: Sci Search, ISI: The ISI Alerting Services, ISI: CompuMath Citation Index, ISI: Current Contents: Engineering, Technology and Applied Science, Geographical Abstracts: Physical Abstracts, INSPEC Information Services, Zentralblatt für Mathematik, Computer Literature Index, Engineering Index, Ei Page One, COMPENDEX Plus, ACM Computing Reviews, ACM Guide to Computing Literature, and Information Science Abstracts.*

Subscription Rates

The subscription price for 1999 Volume 7 (4 quarterly issues) is:
Institutions: US $327.50   NLG 655.00
Individuals: US $154.00   NLG 325.00

The above rates are inclusive of postage and handling. The individual rate is not available to institutions, libraries, or companies. For airmail delivery, please add US $9.50/NLG 22.00. The journal is shipped to the USA and Canada in bulk airfreight at no extra cost.

Second-class postage paid at Rahway, N.J.

U.S. mailing agent: Mercury Airfreight International, Ltd., Inc.
2323 Randolph Ave.
Avenel, N.J. 07001, USA

ISSN: 0926-8782   USPS: 010-328

Postmaster: Please send all address corrections to: *Distributed and Parallel Databases*, c/o Mercury Airfreight International, Ltd., Inc., 2323 Randolph Ave., Avenel, N.J. 07001, USA.

Ordering Information/Sample Copies:

Subscription orders and requests for sample copies should be sent to:

Kluwer Academic Publishers   Kluwer Academic Publishers
Post Office Box 358   Post Office Box 322
Accord Station   or   3300 AH Dordrecht
Hingham, MA 02018-0358, USA   THE NETHERLANDS
FAX: 617-871-6528   FAX: 011-317-833-4254

Subscriptions may also be sent to any subscription agent. Private subscriptions should be sent directly to the publisher at the above addresses.

ISSN 0926-8782

Distributed and Parallel Databases, 7, 255–256 (1999)
© 1999 Kluwer Academic Publishers. Manufactured in The Netherlands.

Guest Editorial

As we approach the turn of the century, mobile devices and wireless data networks are becoming more powerful and affordable. This has lead to the growing importance of mobile data access and management. Mobile technology expands the scope of distributed computing infrastructure and enhances distributed applications by providing nomadic users ubiquitous (anytime, anywhere) access to networked information resources.

However, existing information systems and distributed applications often make inherent assumptions about the connectivity and topology of the underlying networks, as well as the capability of client devices, that are not valid in mobile environments. These assumptions, which are well understood by the research community, present significant difficulty in meeting nomadic users' expectations of having applications operate seamlessly across wireless and wired networks. To address these limitations, researchers have concentrated on creating adaptive systems for mobile applications which can react to the changing mobile environment and allow their functionality to be dynamically customized for the best performance possible.

The research effort has come from both the data management and networking communities, and has particularly focused on new paradigms and architectures for mobile data access and management. Experiences with the design and implementation of applications and computing services supporting mobile data access have also been reported in literature. The aim of this special issue is to present a broad and illustrative picture of this research effort, as well as report on some new developments. To this end, we present four technical papers that cover a range of issues—from new computing models and adaptation paradigms to database applications.

In the paper titled "Updating and Querying Databases that Track Mobile Units", Ouri Wolfson and colleagues describe databases representing information about moving objects (e.g., vehicles), particularly their location. The problems of updating and querying such databases are addressed. Specifically, the update problem is to determine when the location of a moving object in the database (namely its database location) should be updated. This question is addressed by an information cost model that captures uncertainty, deviation, and communication. The DBMS extensions to supporting the databases are also discussed.

The paper titled "Mobility and Extensibility in the StratOSphere Framework" by Daniel Wu, Divyakant Agrawal, and Amr El Abbadi describes the design and implementation of the StratOSphere project, a framework which unifies distributed objects and mobile code applications. This work represents a new computing and object model for code and data resource management in a dynamic network environment. Particularly, the mobility and extensibility in the framework provide a new way to allow applications and systems to adapt to the changing environment for information access.

To support mobile data access and applications, adaptation paradigms have extensively been explored recently for the construction of applications and system supporting. Friday et al. present an asynchronous adaptation paradigm to support mobile applications in their paper titled "Experiences of Using Generative Communications To Support Adaptive Mobile Applications". The authors present an asynchronous distributed systems platform based on the tuple space paradigm in mobile environments. They discuss the experiences of developing and using this platform and the benefits of the tuple space approach for the design and implementation of adaptive mobile applications and information access.

In the fourth paper titled "MODEC: A Multi-granularity Mobile Object-oriented Database Caching Mechanism, Prototype and Performance" by Chen et al., a caching mechanism for object-oriented database systems in a mobile environment called MODEC is described. The goal of MODEC mechanism is to provide a feasible solution to the mobile data access problem under the constraints in a mobile environment. The authors present the implementation of a prototype for MODEC and the empirical performance results obtained from experiments with data from a real-life database.

With rapid advances in wireless communication and networking, and the anticipated deployment of 4G wirless systems, it is our opinion that mobile data access and management will become an increasingly important research area. We hope that the papers in this issue will help covince you that there are hard problems to be solved if data/information centered applications are to take advantage of the new capabilities that wireless systems will offer, and that these papers present interesting (initial) solutions for some of them.

Jin Jing
Anupam Joshi

Distributed and Parallel Databases, 7, 257–287 (1999)
© 1999 Kluwer Academic Publishers. Manufactured in The Netherlands.

Updating and Querying Databases that Track Mobile Units

OURI WOLFSON*   wolfson@eecs.uic.edu
A. PRASAD SISTLA[†]   sistla@eecs.uic.edu
Department of Electrical Engineering and Computer Science, University of Illinois, Chicago, IL 60607

SAM CHAMBERLAIN
Army Research Laboratory, Aberdeen Proving Ground, MD

YELENA YESHA
Center of Excellence in Space Data and Information Sciences at NASA, Goddard Space Flight Center, Greenbelt, MD

Abstract.   In this paper, we consider databases representing information about moving objects (e.g., vehicles), particularly their location. We address the problems of updating and querying such databases. Specifically, the update problem is to determine when the location of a moving object in the database (namely its database location) should be updated. We answer this question by proposing an information cost model that captures uncertainty, deviation, and communication. Then we analyze dead-reckoning policies, namely policies that update the database location whenever the distance between the actual location and the database location exceeds a given threshold, x. Dead-reckoning is the prevalent approach in military applications, and our cost model enables us to determine the threshold x. We propose several dead-reckoning policies and we compare their performance by simulation.

Then we consider the problem of processing range queries in the database. An example of a range query is 'retrieve the objects that are currently inside a given polygon P'. We propose a probabilistic approach to solve the problem. Namely, the DBMS will answer such a query with a set of objects, each of which is associated with a probability that the object is inside P.

1.   Introduction

1.1.   Background

Consider a database that represents information about moving objects and their location. For example, for a database representing the location of taxi-cabs a typical query may be: retrieve the free cabs that are currently within 1 mile of 33 N. Michigan Ave., Chicago (to pick-up a customer); or for a trucking company database a typical query may be: retrieve the trucks that are currently within 1 mile of truck ABT312 (which needs assistance); or for a database representing the current location of objects in a battlefield a typical query may be:

---

*Wolfson's research was supported in part by NSF grants IRI-9712967, CCR-9816633, and CCR-9803974, DARPA grant N66001-97-2-8901, Army Research Labs grant DAAL01-96-2-0003, NATO grant CRG-960648.
[†]Sistla's work was supported in part by NSF grants IRI-97-11925 and CCR-9803974.

retrieve the friendly helicopters that are in a given region, or, retrieve the friendly helicopters that are expected to enter the region within the next 10 minutes. The queries may originate from the moving objects, or from stationary users. We will refer to the above applications as MOtion-Database (MOD) applications or moving-objects-database applications.

In the military, MOD applications arise in the context of the digital battlefield (see [22, 23]), and in the civilian industry they arise in transportation systems. For example, Omni-tracs developed by Qualcomm (see [19]) is a commercial system used by the transportation industry, which enables MOD functionality. It provides location management by connecting vehicles (e.g., trucks), via satellites, to company databases.

Currently, MOD applications are being developed in an ad hoc fashion. Database man-agement system (DBMS) technology provides a potential foundation for MOD applications, however, DBMSs are currently not used for this purpose. The reason is that there is a critical set of capabilities that have to be integrated, adapted, and built on top of existing DBMSs in order to support moving objects databases. The added capabilities include, among other things, support for spatial and temporal information, support for rapidly changing real-time data, new indexing methods, and imprecision management. The objective of our Databases fOr MovINg Objects (DOMINO) project is to build an envelope containing these capabili-ties on top of existing DBMSs.

In this paper, we address the imprecision problem. The location of a moving object is inherently imprecise because, regardless of the policy used to update the database location of a moving object (i.e., the object-location stored in the database), the database location cannot always be identical to the actual location of the object. There may be several location update policies, for example, the location is updated every x time units. In this paper, we address dead-reckoning policies, namely policies that update the database whenever the distance between the actual location of a moving object m and its database location exceeds a given threshold h, say 1 mile. This means that the DBMS will answer a query "what is the current location of m?" by an answer A: "the current location is (x, y) with a deviation of at most 1 mile". Dead-reckoning is the prevalent approach in military applications.

One of the main issues addressed in this paper is how to determine the update threshold h in dead-reckoning policies. This threshold determines the location imprecision, which encompasses two related but different concepts, namely deviation and uncertainty. The deviation of a moving object m at a particular point in time t is the distance between m's actual location at time t, and its database location at time t. For the answer A above, the deviation is the distance between the actual location of m and (x, y). On the other hand, the uncertainty of a moving object m at a particular point in time t is the size of the area in which the object can possibly be. For the answer A above, the uncertainty is the area of a circle with radius 1 mile. The deviation has a cost (or penalty) in terms of incorrect decision making, and so does the uncertainty. The deviation (resp. uncertainty) cost is proportional to the size of the deviation (resp. uncertainty). The ratio between the costs of an uncertainty unit and a deviation unit depends on the interpretation of an answer such as A above, as will be explained in Section 3.

In MOD applications the database updates are usually generated by the moving objects themselves. Each moving object is equipped with a Geographic Positioning System (GPS), and it updates its database location using a wireless network (e.g., ARDIS, RAM Mobile

4

Data Co., IRIDIUM, etc.). This introduces a third information cost component, namely communication. For example, RAM Mobile Data Co. charges a minimum of 4 cents per message, with the exact cost depending on the size of the message. Furthermore, there is a tradeoff between communication and imprecision in the sense that the higher the communication cost the lower the imprecision and vice versa. In this paper we propose a model of the information cost in moving objects databases, which captures imprecision and communication. The tradeoff is captured in the model by the relative costs of an uncertainty unit, a deviation unit, and a communication unit.

1.2.   Location update policies

Consider an object m moving along a prespecified route. We model the database location of m by storing in the database m's starting time, starting location, and a prediction of future locations of the object. In this paper the prediction is given as the speed v of the object. Thus the database location of m can be computed by the DBMS at any subsequent point in time.[1] This method of modeling the database location was originally introduced in [11, 12] via the concept of a dynamic attribute; the method is modified here in order to handle uncertainty. The actual location of a moving object m deviates from its database location due to the fact that m does not travel at the constant speed v.

A *dead-reckoning update policy* for m dictates that there is a database-update threshold th, i.e., a deviation for which m should send a location/speed update to the database. (Note that at any point in time, since m knows its actual location and its database location, it can compute its current deviation.) *Speed dead-reckoning*[2] (sdr) is a dead-reckoning policy in which the threshold th is fixed for the duration of the trip.

In this paper we introduce another dead-reckoning update policy, called *adaptive dead reckoning(adr)*. Adr provides with each update a new threshold th that is computed using a cost based approach. th minimizes the total information cost per time unit until the next update. The total information cost consists of the update cost, the deviation cost, and the uncertainty cost. In order to minimize the total information cost per time unit between now and the next update, the moving object m has to estimate when the next update will occur, i.e., when the deviation will reach the threshold. Thus, at location update time, in order to compute the new threshold, adr predicts the future behavior of the deviation. The thresholds differ from update to update because the predicted behavior of the deviation is different.

A problem common to both sdr and adr is that the moving object may be disconnected or otherwise unable to generate location updates. In other words, although the DBMS "thinks" that updates are not generated since the deviation does not exceed the update threshold, the actual reason is that the moving object is disconnected. To cope with this problem we introduce a third policy, "disconnection detecting dead-reckoning (dtdr)". The policy avoids the regular process of checking for disconnection by trying to communicate with the moving object, thus increasing the load on the low bandwidth wireless channel. Instead, it uses a novel technique that decreases the uncertainty threshold for disconnection detection. Thus, in dtdr the threshold continuously decreases as the time interval since the last location update increases. It has a value K during the first time unit after the update, it has value $K/2$ during the second time unit after the update, it has value $K/3$ during the third time unit, etc. Thus, if the object is connected, it is increasingly likely that it will generate an update. Conversely, if the moving object does not generate an update, as the time interval since the last update increases it is increasingly likely that the moving object is disconnected. The dtdr policy computes the K that minimizes the total information cost, i.e., the sum of the update cost, the deviation cost, and the uncertainty cost.

To contrast the three policies, observe that for sdr the threshold is fixed for all location updates. For adr the threshold is fixed between each pair of consecutive updates, but it may change from pair to pair. For dtdr the threshold decreases as the period of time between a pair of consecutive updates increases.

We compare by simulation the three policies introduced in this paper namely adr, dtdr, and sdr. The parameters of the simulation are the following. The update-unit cost, namely the cost of a location-update message; the uncertainty-unit cost, namely the cost of a unit of uncertainty; deviation-unit cost, namely the cost of a unit of deviation; a speed curve, namely a function that for a period of time gives the speed of the moving object at any point in time. The comparison is done by quantifying the total information cost of each policy for a large number of combinations of the parameters. Our simulations indicate that adr is superior to sdr in the sense that it has a lower or equal information cost for every value of the update-unit cost, uncertainty-unit cost, and deviation-unit cost. Adr is superior to dtdr in the same sense; the difference between the costs of the two policies quantifies the cost of disconnection detection. For some parameters combinations the information cost of sdr is six times as high as that of adr.

Additionally, we compare the above policies with the best update policy developed in our previous work [12], called immediate-linear (il). The policies in [12] are not dead-reckoning policies, i.e., they do not update the database when the deviation reaches some bound known to the DBMS. Instead, the update time-point depends on the overall behavior of the deviation since the last update. Partly as a consequence of this, the il policy cannot consider the uncertainty factor in deciding when to update the database. Although the moving object does not provide an uncertainty at location-update time, the DBMS can in some cases compute an upper limit on the uncertainty using additional assumptions such as maximum and minimum speed of the object. However, these upper limits are unnecessarily high. Thus, when considering the cost of uncertainty the total information cost of the il policy is higher than the cost of adr, and often higher than the cost of dtdr as well.

Finally, an additional contribution of this paper is a probabilistic model and an algorithm for query processing in motion databases. In our model the location of the moving object is a random variable, and at any point in time the database location and the uncertainty are used to determine a density function for this variable. Based on this model we developed an algorithm that processes range queries such as Q = 'retrieve the moving objects that are currently inside a given region R'. The answer to Q is a set of objects, each of which is associated with the probability that currently the object is inside R.

In summary, our contributions in this paper are as follows:

- We propose an information cost model for moving objects databases that captures uncertainty, deviation and communication.

- We propose the adr database update policy for moving objects. It adapts the dead-reckoning threshold to the relative costs of the deviation, uncertainty and communication, and to the predicted behavior of the deviation, so that the total information cost is minimized.
- We introduce a novel technique for disconnection detection in mobile computing, namely decreasing uncertainty threshold. Intuitively, it postulates that the probability of communication should increase as the period of time since the last communication increases. Based on the decreasing uncertainty threshold technique, we propose the dtdr database update policy for moving objects. The initial threshold is optimized to minimize the total information cost.
- We use a simulation testbed to compare the policies, sdr, adr, dtdr, and the il policy developed in previous work [12]. We conclude that adr is superior to all the other policies in the sense that it has a lower information cost. The difference between the cost of dtdr and that of adr is the cost of disconnection detection.
- We introduce a probabilistic model and algorithm for processing range queries in motion databases.

The rest of this paper is organized as follows. In Section 2 we introduce the data model and discuss location attributes of moving objects. In Section 3 we discuss the information cost of a trip, and in Section 4 we introduce our approach to cost optimization. In Section 5 we describe the three location update policies. In Section 6 we present our approach to probabilistic query processing. In Section 7 we discuss relevant work, and in the last section we conclude and discuss future work. In Appendix A we discuss the comparison of the update policies by simulation. In Appendix B we demonstrate the adr and dtdr update policies by examples.

2.   The data model

In this section, we define the main concepts used in this paper. A *database* is a set of object-classes. An *object-class* is a set of attributes. Some object-classes are designated as *spatial*. Each spatial object class is either a point-class, a line-class, or a polygon-class in two-dimensional space (all our concepts and results can be extended to three-dimensional space).

Point object classes are either mobile or stationary. A point object class O has a *location attribute L*. If the object class is *stationary*, its location attribute has two sub-attributes $L.x$, and $L.y$, representing the x and y coordinates of the object. If the object class is *mobile*, its location attribute has six sub-attributes, *L.route, L.startlocation, L.starttime, L.direction, L.speed*, and *L.uncertainty*.

The semantics of the sub-attributes are as follows. *L.route* is (the pointer to) a line spatial object indicating the route on which an object in the class O is moving. Although we assume that the objects move along predefined routes, our results can be extended to free movement in space (e.g., by aircraft). We will comment on that option in the last paragraph of this section. *L.startlocation* is a point on *L.route*; it is the location of the moving object at time

L.starttime. In other words, *L.starttime* is the time when the moving object was at location *L.startlocation.* We assume that whenever a moving object updates its *L* attribute it updates the *L.startlocation* subattribute; thus at any point in time *L.starttime* is also the time of the last location-update. We assume in this paper that the database updates are instantaneous, i.e., valid- and transaction-times (see [21]) are equal. Therefore, *L.starttime* is the time at which the update occurred in the real world system being modeled, and also the time when the database installs the update. *L.direction* is a binary indicator having a value 0 or 1 (these values may correspond to north-south, or east-west, or the two endpoints of the route). *L.speed* is a function that represents the predicted future locations of the object. It gives the distance of the moving object from *L.startlocation* as a function of the number t of time units elapsed since the last location-update, namely since *L.starttime.* The function has the value 0 when $t = 0$. In its simplest form (which is the only form we consider in this extended abstract) *L.speed* represents a constant speed v, i.e., the distance is $v \cdot t$.[3] *L.uncertainty* is either a constant, or a function of the number t of time units elapsed since *L.starttime.* It represents the threshold on the location deviation (the deviation is formally defined at the end of this section); when the deviation reaches the threshold, the moving object sends a location update message. Observe that the uncertainty may change automatically as the time elapsed since *L.starttime* increases; this is indeed the case for the dtdr policy.

We define the *route-distance* between two points on a given route to be the distance along the route between the two points. We assume that it is straightforward to compute the route-distance between two points, and the point at a given route-distance from another point. The *database location* of a moving object at a given point in time is defined as follows. At time *L.starttime* the database location is *L.startlocation*; the database location at time *A.starttime* $+ t$ is the point (x, y) which is at route-distance *L.speed* $\cdot t$ from the point *L.startlocation.* Intuitively, the database location of a moving object m at a given time point t is the location of m as far as the DBMS knows; it is the location that is returned by the DBMS in response to a query entered at time t that retrieves m's location. Such a query also returns the uncertainty at time t, i.e., it returns an answer of the form: m is on *L.route* at most *L.uncertainty* ahead of or behind (x, y).

Since between two consecutive location updates the moving object does not travel at exactly the speed *L.speed*, the actual location of the moving object deviates from its database location. Formally, for a moving object, the *deviation d* at a point in time t, denoted $d(t)$, is the route-distance between the moving object's actual location at time t and its database location at time t. The deviation is always nonnegative. At any point in time the moving object knows its current location, and it also knows all the subattributes of its location attribute. Therefore, at any point in time the (computer onboard the) moving object can compute the current deviation. Observe that at time *L.starttime* the deviation is zero.

At the beginning of the trip the moving object updates all the sub-attributes of its location attribute. Subsequently, the moving object periodically updates its current location and speed stored in the database. Specifically, a *location update* is a message sent by the moving object to the database to update some or all the sub-attributes of its location attribute. The moving object sends the location update when the deviation exceeds the *L.uncertainty* threshold, or when the moving object changes route or direction. The location update message contains at least the values for *L.speed* and *L.startlocation*. Obviously, other subattributes can also be updated. The subattribute *L.starttime* is written by the DBMS whenever it installs a location update; it denotes the time when the installation is done.

Before concluding this section we would like to point out that the results of this paper hold for free-movement modeling, i.e., for objects that move freely in space (e.g., aircraft) rather than on routes. In this case, *L.route* is an infinite straight line (e.g., 60° from the starting point) rather than a line-object stored in the database. Then there are two possibilities of modeling the uncertainty. The first is identical to the one described above, i.e., the uncertainty is a segment on the infinite line representing the route. In this case, every change of direction constitutes a change of route, thus necessitating a location update. The second possibility is to redefine the deviation to be the Euclidean distance between the database location and the actual location, and to remove the requirement that the object updates the database whenever it changes routes. In this case, *L.uncertainty* defines a circle around the database location, and a query that retrieves the location of a moving object *m* returns an answer of the form: *m* is within a circle having a radius of at most *L.uncertainty* from (x, y). Observe that the second possibility of modeling uncertainty necessitates less location updates, but the answer to a query is less informative since the uncertainty is given in two-dimensional space rather than one-dimensional.

3. The information cost of a trip

In this section, we define the information cost model for a trip taken by a moving object *m*, and we discuss information cost optimality.

At each point in time during the trip the moving object has a deviation and an uncertainty, each of which carries a penalty. Additionally, the moving object sends location update messages. Thus the information cost of a trip consists of the cost of deviation, cost of communication, and cost of uncertainty.

Now we define the deviation cost. Observe first that the cost of the deviation depends both on the size of the deviation and on the length of time for which it persists. It depends on the size of the deviation since decision-making is clearly affected by it. To see that it depends on the length of time for which the deviation persists, suppose that there is one query per time unit that retrieves the location of a moving object *m*. Then, if the deviation persists for two time units its cost will be twice the cost of the deviation that persists for a single time unit; the reason is that two queries (instead of one) will pay the deviation penalty. Formally, for a moving object *m* the cost of the deviation between two time points t_1 and t_2 is given by the *deviation cost function*, denoted $COST_d(t_1, t_2)$; it is a function of two variables that maps the deviation between the time points t_1 and t_2 into a nonnegative number. In this paper we take the penalty for each unit of deviation during a unit of time to be one (1). Then, the cost of the deviation between two time points t_1 and t_2 is:

$$COST_d(t_1, t_2) = \int_{t_1}^{t_2} d(t) \, dt \qquad (1)$$

(Recall that $d(t)$ is the deviation as a function of time).

9

The *update cost*, denoted C_1, is a nonnegative number representing the cost of a location-update message sent from the moving object to the database. This is the cost of the resources (i.e., bandwidth and computation) consumed by the update. The update cost may differ from one moving object to another, and it may vary even for a single moving object during a trip, due for example, to changing availability of bandwidth. The update cost must be given in the same units as the deviation cost. In particular, if the update cost is C_1 it means the ratio between the update cost and the cost of a unit of deviation per unit of time (which is one) is C_1. It also means that the moving object (or the system) is willing to use $1/C_1$ messages in order to reduce the deviation by one during one unit of time.

Now we define the uncertainty cost. Observe that, as for the deviation, the cost of the uncertainty depends both, on the size of the uncertainty and on the length of time for which it persists. Formally, for a moving object m the cost of the uncertainty between two time points t_1 and t_2 is given by the *uncertainty cost function*, denoted $COST_u(t_1, t_2)$; it is a function of two variables that maps the uncertainty between the time points t_1 and t_2 into a nonnegative number. Define the *uncertainty unit cost* to be the penalty for each unit of uncertainty during a unit of time, and denote it by C_2. Then, the cost of the uncertainty of m between two time points t_1 and t_2 is:

$$COST_u(t_1, t_2) = \int_{t_1}^{t_2} C_2 u(t)\, dt \tag{2}$$

where $u(t)$ is the value of the *L.uncertainty* subattribute of m as a function of time.

The uncertainty unit cost C_2 is the ratio between the cost of a unit of uncertainty and the cost of a unit of deviation. Consider an answer returned by the DBMS: "the current location of the moving object m is (x, y), with a deviation of at most u units". C_2 should be set higher than 1 if the uncertainty in such an answer is more important than the deviation, and lower than 1 otherwise. Observe that in a dead-reckoning update policy each update message establishes a new uncertainty which is not necessarily lower than the previous one. Thus communication reduces the deviation but not necessarily the uncertainty.

Now we are ready to define the information cost of a trip taken by a moving object m. Let t_1 and t_2 be the time-stamps of two consecutive location update messages. Then the *information cost* in the interval $[t_1, t_2)$ is:

$$COST_I[t_1, t_2) = C_1 + COST_d[t_1, t_2) + COST_u[t_1, t_2) \tag{3}$$

Observe that $COST_I[t_1, t_2)$ includes the message cost at time t_1 but not the cost of the one at time t_2. Observe also that each location update message writes the actual current location of m in the database, thus it reduces the deviation to zero. The total information cost of a trip is computed by summing up $COST_I[t_1, t_2)$ for every pair of consecutive update points t_1 and t_2. Formally, let the time points of the update messages sent by m be $t_1, t_2, \ldots, t_k$. Furthermore, let 0 be the time point when the trip started and t_{k+1} the time point when the trip ended. Then the *total information cost* of a trip is

$$COST_I = COST_d[0, t_1) + COST_u[0, t_1) + \sum_{i=1}^{k} COST_I[t_i, t_{i+1}) \tag{4}$$

10

4. Cost based optimization for dead reckoning policies

As mentioned in the introduction, a dead-reckoning update policy for a moving object m dictates that at any point in time there is a database-update threshold th, of which both the DBMS and m are aware. When the deviation of m reaches th, m sends to the database an update consisting of the current location, the predicted speed, and the new deviation threshold K. The objective of the dead reckoning policies that we introduce in this paper is to set K (which the DBMS installs in the $L.uncertainty$ subattribute), such that the total information cost is minimized. Intuitively, this is done as follows. First, m predicts the future behavior of the deviation. Based on this prediction, the average cost per time time unit between now and the next update is obtained as a function f of the new threshold K. Then K is set to minimize f.[4] It is important to observe that we optimize the average cost per time unit rather than simply the total cost between the two time points; clearly, the total cost increases as the time interval until the next update increases.

The next theorem establishes the optimal value K for $L.uncertainty$ under the assumption that the deviation between two consecutive updates is a linear function of time.

Theorem 1. *Denote the update cost by C_1, and the uncertainty unit cost by C_2. Assume that for a moving object two consecutive location updates occur at time points t_1 and t_2. Assume further that between t_1 and t_2, the deviation $d(t)$ is given by the function $a(t - t_1)$ where $t_1 \leq t \leq t_2$ and a is some positive constant; and $L.uncertainty$ is fixed at K throughout the interval (t_1, t_2). Then the total information cost per time unit between t_1 and t_2 is minimized if $K = \sqrt{2aC_1/(2C_2 + 1)}$.*

Proof: Based on Eqs. (1)–(3):

$$COST_I[t_1, t_2] = C_1 + \int_{t_1}^{t_2} a(t - t_1)\, dt + C_2 K(t_2 - t_1)$$

$$= C_1 + \frac{a(t_2 - t_1)^2}{2} + C_2 K(t_2 - t_1) \tag{5}$$

Denote by $f(t_2)$ the average information cost per time unit between t_1 and t_2, for the update time t_2. Namely,

$$f(t_2) = \frac{COST_I[t_1, t_2]}{(t_2 - t_1)} \tag{6}$$

Since the update immediately following t_1 occurs at time t_2, it means that at that time the deviation reaches the threshold $L.uncertainty$, namely:

$$K = a(t_2 - t_1) \tag{7}$$

Thus we can substitute $K/a + t_1$ for t_2 in Eq. (6) and obtain $f(K) = \frac{aC_1}{K} + (\frac{1}{2} + C_2)K$. Using the derivative it is easy to calculate that the minimum of $f(K)$ is obtained when $K = \sqrt{2aC_1/(1 + 2C_2)}$. □

The implication of Theorem 1 is the following. Suppose that a moving object m is currently at time point t_1, i.e., its deviation has reached the uncertainty threshold $L.uncertainty$. Now m needs to compute a new value for $L.uncertainty$ and send it in the location update message. Suppose further that m predicts that following the update the deviation will behave as the linear function $a(t - t_1)$, and in the update message it has to set the uncertainty threshold $L.uncertainty$ to a value that will remain fixed until the next update. Then, in order to optimize the information cost, m should set the threshold to $K = \sqrt{2aC_1/(2C_2 + 1)}$.

Next assume that, in order to detect disconnection, one is interested in a dead-reckoning policy in which the uncertainty threshold $L.uncertainty$ continuously decreases between updates. Particularly, we consider a particular type of decrease, that we call fractional decrease; other types exist, but we found this one convenient. Let K be a constant. If the uncertainty threshold $L.uncertainty$ *decreases fractionally starting with* K, then during the first time unit after a location update u its value is K, during the second time unit after u its value is $K/2$, during the third time unit after u its value is $K/3$, etc., until the next update (which establishes a new K).

Theorem 2. *Assume that for a moving object two consecutive location updates occur at time points t_1 and t_2. Assume further that between t_1 and t_2, the deviation $d(t)$ is given by the function $a(t - t_1)$ where $t_1 \le t \le t_2$ and a is some positive constant; and in the time interval (t_1, t_2) L.uncertainty decreases fractionally starting with a constant K. Then the total information cost per time unit between t_1 and t_2 is given by the following function of K.*

$$f(K) = \frac{C_1 + \frac{1}{2}K + C_2 K \left(1 + \frac{1}{2} + \frac{1}{3} + \cdots + \frac{1}{\sqrt{K/a}}\right)}{\sqrt{K/a}}.$$

Proof: Based on Eqs. (1)–(3):

$$COST_I[t_1, t_2) == C_1 + \frac{a(t_2 - t_1)^2}{2} + C_2 K \left(1 + \frac{1}{2} + \frac{1}{3} + \cdots + \frac{1}{t_2 - t_1}\right) \tag{8}$$

The deviation at time t_2 is $a(t_2 - t_1)$, and the threshold at that time is $K/(t_2 - t_1)$. Since the deviation reaches the threshold at time t_2 the above values are equal, and therefore $K = a(t_2 - t_1)^2$ and $\sqrt{K/a} = (t_2 - t_1)$. Thus, if we divide Eq. (8) by $(t_2 - t_1)$ in order to obtain the information cost *per time unit*, and then substitute K for $a(t_2 - t_1)^2$ and $\sqrt{K/a}$ for $(t_2 - t_1)$ in the resulting equation, the theorem follows.   □

Similar to Theorem 1, the implication of Theorem 2 is the following. Suppose that a moving object is currently at time point t_1, i.e., it is about to send a location update message, and it can predict that following the update the deviation will behave as the linear function $a(t - t_1)$, and in the update message it sets the uncertainty threshold $L.uncertainty$ to a fractionally decreasing value starting with K. Then in order to optimize the information cost it should set K to the value that minimizes the function of Theorem 2.

12

5. Description of the location update policies

In this section, we describe and motivate the following three location update policies.

The speed dead-reckoning (sdr) policy.   At the beginning of the trip the moving object m sends to the DBMS an uncertainty threshold that is selected in an ad hoc fashion, it is stored in *L.uncertainty*, and it remains fixed for the duration of the trip.   The object m updates the database whenever the deviation exceeds *L.uncertainty*; the update simply includes the current location and current speed.[5]

The adaptive dead reckoning (adr) policy.   At the beginning of the trip the moving object m sends to the DBMS an initial deviation threshold th_1 selected arbitrarily.[6] Then m starts tracking the deviation. When the deviation reaches th_1, the moving object sends an update message to the database. The update consists of the current speed, current location, and a new threshold th_2 that the DBMS should install in the *L.uncertainty* subattribute. th_2 is computed as follows. Denote by t_1 the number of time units from the beginning of the trip until the deviation reaches th_1 for the first time, by I_1 the cost of the deviation (which is computed using Eq. (1)) during the same time interval, and let $a_1 = \frac{2I_1}{t_1^2}$. Then th_2 is $\sqrt{2a_1 C_1/(1 + 2C_2)}$ (remember, C_1 is the update cost, C_2 is the unit-uncertainty cost). When the deviation reaches th_2, a similar update is sent, except that the new threshold th_3 is $\sqrt{2a_2 C_1/(1 + 2C_2)}$, where $a_2 = \frac{2I_2}{t_2^2}$ (I_2 is the cost of the deviation from the first update to second update, t_2 is the number of time units elapsed since the first location update). Since a_2 may be different than a_1, th_2 may be different than th_3. When th_3 is reached the object will send another update containing th_4 (which is computed in a similar fashion), and so on.

The mathematical motivation for adr is based on Theorem 1 in a straightforward way. Namely, at each update time point p_i adr simply sets the next threshold in a way that optimizes the information cost per time unit (according to Theorem 1), assuming that the deviation following time p_i will behave as the following linear function: $d(t) = \frac{2I_i}{t_i^2}t$, where t is the number of time units after p_i, and t_i is the number of time units between the immediately preceding update and the current one (at time p_i), and I_i the cost of the deviation during the same time interval. The reason for this prediction of the future deviation is as follows. Adr approximates the current deviation, i.e., the deviation from the time of the immediately preceding update to time p_i, by a linear function[7] (see [24]) with slope $\frac{2I_i}{t_i^2}t$. Observe that at time p_i this linear function has the same deviation cost (namely I_i) as the actual current deviation.[8] Based on the locality principle, adr predicts that after the update at time p_i, the deviation will behave according to the same approximation function.

For example, consider figure 1. At time point p_i adr predicts that the future deviation is given by the linear function with slope l.

The disconnection detection dead reckoning (dtdr) policy.   At the beginning of the trip the moving object m sends to the DBMS an initial deviation threshold th_1 selected arbitrarily. The moving objects sets the uncertainty threshold *L.uncertainty* to a fractionally decreasing value starting with th_1. That is, during the first time unit the uncertainty threshold is th_1;

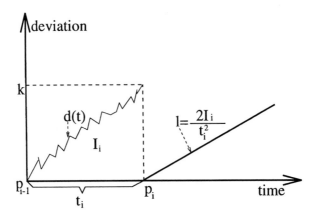

Figure 1. At time p_i the deviation is predicted to behave as a linear function with slope l.

during the second time unit period it is $\frac{th_1}{2}$, and so on. Then it starts tracking the deviation. At time t_1 when the deviation reaches the current uncertainty threshold, namely $\frac{th_1}{t_1}$, the moving object sends a location update message to the database. The update consists of the current speed, current location, and a new threshold th_2 to be installed in the *L.uncertainty* subattribute.

th_2 is computed using the function $f(K)$ of Theorem 2. Since $f(K)$ uses the slope a of the future deviation, we first estimate the future deviation as in the adr case, as follows. Denote by I_1 the cost of the deviation (which is computed using Eq. (1)) since the beginning of the trip, and let $a_1 = \frac{2I_1}{t_1^2}$. Now, observe that $f(K)$ does not have a closed form formula. Thus, we first approximate the sum $\frac{1}{2} + \frac{1}{3} + \cdots + \frac{1}{\sqrt{K/a_1}}$ by $\ln(\sqrt{K/a_1})$ (since $\ln n$ is an approximation for the nth harmonic number). Thus, the approximation function for $f(K)$ is $g(K) = \frac{C_1 + \frac{K}{2} + C_2 K (\ln(\sqrt{K/a_1})+1)}{\sqrt{K/a_1}}$. The derivative of $g(K)$ is zero when K is the solution to the following equation.

$$\ln(K) = \frac{d_1}{K} - d_2 \tag{9}$$

where in the equation, $d_1 = \frac{2C_1}{C_2}$ and $d_2 = \frac{1}{C_2} + 4 - \ln(a_1)$. We find a numerical solution to this equation using the Newton Raphson method. The solution is the new threshold th_2, and the moving object sets the uncertainty threshold *L.uncertainty* to a fractionally decreasing value starting with th_2.

After t_2 time units, when the deviation reaches the current uncertainty threshold, namely $\frac{th_2}{t_2}$, a location update containing th_3 is sent. th_3 is computed as above, except that a new slope (as in adr) a_2 is used; and I_2 is the cost of the deviation during the previous t_2 time units. The process continues until the end of the trip. That is, at each update time point, dtdr determines the next optimal threshold by the constants C_1, C_2, and the slope a_i of the current deviation approximation function.

14

6. Querying with uncertainty

In this section, we present a probabilistic method for specifying and processing range queries about motion databases. For example, a typical query might be "Retrieve all objects o which are within the region R". Since there is an uncertainty about the location of the various objects at any time, we may not be able to answer the above query with absolute certainty. Instead, our query processing algorithm outputs a set of pairs of the form (o, p) where o is an object and p is the probability that the object is in region R at time t; actually, the algorithm retrieves only those pairs for which p is greater than some minimum value. Note that here we are using probability as a measure of certainty.

As indicated, we assume that all the objects are traveling on routes. Since the actual location is not exactly known, we assume that the location of an object o on its route at time t is a random variable [8]. We let $f_o(x)$ denote the density function of this random variable. More specifically, for small values of dx, $f_o(x)dx$ denotes the probability that o is at some point in the interval $[x, x + dx]$ at time t (actually, f_o is a function of x and t; however we omit this as t is understood from the context). The mean m_o of the above random variable is given by the database location of o (this equals $o.L.startlocation + o.L.speed(t - o.L.starttime)$; see Section 2).

Now we discuss some possible candidates for the density functions f_o. Many natural processes tend to behave according to the normal density function. Let $\mathcal{N}_{m,\sigma}(x)$ denote a normal density function with mean m and standard deviation σ. We can adopt the normal density functions follows. We take the mean m to be equal to m_o given in the previous paragraph. Next, we relate the standard deviation to the uncertainty of the object location. We do this by setting $\sigma = \frac{1}{c}(o.L.uncertainty)$ where $c > 0$ is constant. In this case, the probability that the object is within a distance of $o.L.uncertainty$ (i.e., within a distance of $c\sigma$) from the location m_o will be higher for higher values of c; for example, this probability will be equal to .68, .95 and .997 for values of c equal to 1, 2 and 3, respectively (see [8]). The value of c is a function of the update policy used, the reliability of the network, the time since the last update and the ratio between uncertainty and deviation unit costs. Whatever may be the value c, there is still a non-zero probability p that the object is at a distance greater than $o.L.uncertainty$ from the the mean m_o; we can interpret this probability to be the probability that there is a disconnection. An alternative is to make p zero. This can be done by modifying the normal distribution to be a bounded normal distribution. This is done by conditioning that the object is within distance $o.L.uncertainty$ from m_o. More specifically, we first use the normal distribution as above by choosing an appropriate c. Then we compute the probability q that the object is within a distance of $o.L.uncertainty$ from m_o. Define the density function $f_o(x)$ to be equal to $\frac{1}{q}\mathcal{N}_{m_o,\sigma}(x)$ for values of x within the interval $[(m_o - o.L.uncertainty), (m_o + o.L.uncertainty)]$ and to be zero for other values of x.

A range query on motion databases is of the following form:

RETRIEVE o FROM Moving-objects WHERE C.

We assume that the condition part C, in the above query, is a conjunction of two conditions p and q; p only refers to the static attributes of the objects (such as 'type', 'color' etc.) and is called the static part; q depends upon the location attributes and is called the dynamic part of the condition. We assume that there is only one object variable that appears free in C, and hence in p and q. Now we describe a method to process the query. Using the underlying database management system we execute the query whose condition part is only p. The set of objects thus retrieved are processed against the condition q and appropriate probability values are calculated as follows.

We assume that the dynamic part of the query condition is formed by using atomic predicates $inside(o, R)$, $within\_distance(o, R, d)$ and boolean connectives $\wedge$ ("and") and $\neg$ ("not")(note that $\vee$, i.e., the "or" operator, can be defined using $\wedge$ and $\neg$). In the atomic predicate $inside(o, R)$, o is an object variable and R is the name of a region. An object o_1 satisfies this predicate at time t if its location lies within the region R. The predicate $within\_distance(o, R, d)$ is satisfied by an object o_1 traveling on route r_1 if the location of o_1 is within distance d of the region R (here the distance is measured along the route, i.e., the route distance). The following is an example query.

RETRIEVE o FROM Moving-objects WHERE $o.type =' ambulance' \wedge inside(o, R)$

Consider an object o_1 traveling on route r_1. We assume that the route r_1 intersects the region R at different places, and certain segments of the route are in the region R; each such segment is given by an interval $[u, v]$ (where $u \leq v$). For the route r_1, we let $Inside\_Int(r_1, R)$ denote the set of all such intervals. Clearly, object o_1 is in region R at time t, if its location at t lies within any of the intervals belonging to $Inside\_Int(r_1, R)$. Using the set of intervals $Inside\_Int(r_1, R)$, we can easily compute another set of intervals on route r_1, denoted by $Within\_Int(r_1, R, d)$, such that every point belonging to any of these intervals is within distance d of region R.

Now consider a condition q formed using the above atomic predicates and using the boolean connectives. As indicated earlier, we assume that q has only one free object variable o. Now we describe a procedure for evaluation of this condition against a set of objects. The satisfaction of this condition by an object o_1 traveling on route r_1 at time t only depends on the location of the object at time t. We first compute the set of all such points. We say that a point x on the route r_1 satisfies the query q, if an object o_1 at location x satisfies q. By a simple induction on the length of q, it is easily seen that the set of points on route r_1 that satisfy q is given by a collection of disjoint intervals (if q is an atomic predicate then this is trivially the case as indicated earlier; for a more complex query, later we give an algorithm for generating this set of intervals). We let $Int(r_1, q)$ denote this set of intervals.

Theorem 3. *For a condition q and route r_1, let $\{I_1, \ldots, I_i, \ldots, I_k\}$ be all the intervals in $Int(r_1, q)$, where $I_i = [u_i, v_i]$. Then, the probability that object o_1 traveling on route r_1 satisfies q at time t is given by*

$$\sum_{i=1}^{k} \int_{u_i}^{v_i} f_{o_1}(x)\, dx$$

16

Proof: The probability that object o_1 satisfies q at time t equals the probability that the current location of o_1 lies within any of the intervals in $Int(r_1, q)$. Let $\{I_1, I_2, \ldots, I_k\}$ be all the intervals in $Int(r_1, q)$. Since all the intervals in $Int(r_1, q)$ are disjoint, it is the case that for any two distinct intervals I_i and I_j the events indicating that o_1 is inside the interval I_i (resp., inside I_j) are independent. Hence, the probability that o_1 satisfies q is equal to the sum, over all intervals I in $Int(r_1, q)$, of the probability that o_1 is in the interval I. For any interval $I_i = [u_i, v_i]$, the probability that the location of o_1 is in the interval I_i equals $\int_{u_i}^{v_i} f_{o_1}(x)\, dx$. The theorem follows from this and our earlier observations.   □

A simple algorithm for computing $Int(r_1, q)$ is given below. For the route r_1, the set of intervals $Int(r_1, q)$ is computed inductively on the structure of q as follows:

q **is an atomic predicate:** If q is $inside(o, R)$, $Int(r_1, q)$ is the same as $Inside\_Int(r_1, R)$ and this is obtained directly from the database, possibly using a spatial indexing scheme. If q is $within\_distance(o, R, d)$ then $Int(r_1, q)$ is same as $Within\_Int(r_1, R, d)$, and this can be computed directly from $Inside\_Int(r_1, R)$. The list of intervals $Int(r_1, R)$ is output in sorted order of the first coordinate of the interval.

$q = q_1 \wedge q_2$**:** First we compute the lists $Int(r_1, q_1)$ and $Int(r_1, q_2)$. After this, we take an interval I_1 from the first list and an interval I_2 from the second list, and output the interval $I_1 \cap I_2$ (if it is non-empty); the set of all such intervals will be the output. Since the original two lists are sorted, the above procedure can be implemented by a modified merge algorithm. The complexity of this procedure is proportional to the sum of the two input lists.

$q = \neg q_1$**:** First we compute $Int(r_1, q_1)$. We assume that the length of the route r_1 is l_1; thus the set of all points on r_1 is given by the single interval $[0, l_1]$. The set of all points on r_1 that satisfy q is the complement of the set of points that satisfy q_1 where this complement is taken with respect to all the points on the route; clearly, this set of points is a collection of disjoint intervals. Now, it is fairly straightforward to see how the sorted list of intervals in $Int(r_1, q)$ can be computed from $Int(r_1, q_1)$; the complexity of such a procedure is simply linear in the number of intervals in $Int(r_1, q_1)$.

If $L_1, L_2, \ldots, L_k$ are the lists of intervals corresponding to the atomic predicates appearing in q and l is the sum of the lengths of these lists, and m is the length of q then it can be shown that the complexity of the above procedure is $O(lm)$.

Now consider the query

RETRIEVE o FROM Moving-objects WHERE $p \wedge q$

where p, q, respectively, are the static and the dynamic parts of the condition.

The overall algorithm for processing the query is as follows:

1. Using the underlying database process the following query:

RETRIEVE o FROM Moving-objects WHERE p.

Let O be the set of objects retrieved.

2. Using the underlying database retrieve the set of routes R on which the objects in O are traveling.
3. For each atomic predicate x appearing in q and for each route r_1 in R, retrieve the list of intervals $Int(r_1, x)$. This is achieved by using any spatial indexing scheme.
4. Using the algorithm presented earlier, for each route r_1, compute the list of intervals $Int(r_1, q)$.
5. For each route r_1 and for each object o_1 traveling on r_1, compute the probability that it satisfies q using the formula given in Theorem 3.

7. Relevant work

One research area to which this paper is related is uncertainty and incomplete information in databases (see, for example, [1, 18] for surveys). However, as far as we know this area has so far addressed complementary issues to the ones in this paper. Our current work on location update policies addresses the question: what uncertainty to initially associate with the location of each moving object. In contrast, existing works are concerned with management and reasoning with uncertainty, after such uncertainty is introduced in the database. Our probabilistic query processing approach is also concerned with this problem. However, our uncertainty processing problem is combined with a temporal-spatial aspect that has not been studied previously as far as we know; for treatment of the spatial issue alone see [9].

Our problem is also related to mobile computing, particularly works on location management in the cellular architecture. These works address the following problem. When calling or sending a message to a mobile user, the network infrastructure must locate the cell in which the user is currently located. The network uses the location database that gives the current cell of each mobile user. The record is updated when the user moves from one cell to another, and it is read when the user is called. Existing works on location management (see, for example, [2, 4, 10, 14, 15, 17, 25]) address the problem of allocating and distributing the location database such that the lookup time and update overhead are minimized. Location management in the cellular architecture can be viewed as addressing the problem of providing uncertainty bounds for each mobile user. The geographic bounds of the cell constitute the uncertainty bounds for the user. Uncertainty at the cell-granularity is sufficient for the purpose of calling a mobile user or sending him/her a message. When it is also sufficient for MOD applications, the location database can be sold by wireless communication vendors to mobile fleet operators. However, often uncertainty at the cell granularity is insufficient. For example, in satellite networks the diameter of a cell ranges from hundreds to thousands of miles.

Another relevant research area is constraint databases (see [16] for a survey and [6, 13] for some notable systems). In this sense, our location attributes can be viewed as a constraint, or a generalized tuple, such that the tuples satisfying the constraint are considered to be in the database. Constraint databases have been separately applied to the temporal (see [3, 5, 7]) domain, and to the spatial domain (see [20]). Constraint databases can be used a framework in which to implement the proposed update policies and query processing algorithm. Since our main problem is communication tradeoffs in replicating

18

the location information, data replication (e.g., [26]) is also somewhat related to our current research.

Finally, the present paper extends the work on which we initially reported in [11, 12] in two important ways. First, in this paper we introduce a quantitative new probabilistic model and method of processing range queries. In contrast, in previous works we took a qualitative approach in the form of "may" and "must" semantics of queries. Second, in this paper we introduce uncertainty as a separate concept from deviation. The previous work on update policies (i.e., [12]) is not equipped to distinguish between uncertainty and deviation. Consequently, The location update policies discussed in this paper are different in two respects from the update policies in [12]. First, they take uncertainty into consideration when determining when to send a location update message. Second they are dead reckoning policies; namely they provide the uncertainty, i.e., the bound on the deviation, with each location update message. In contrast, the [12] policies are not dead reckoning in the sense that the moving object does not update its location when the deviation reaches some threshold; the update time-point depends on the overall behavior of the deviation since the last update. Our simulation results reported in the appendix indicate that the [12] policies are inferior to adr (and often to dtdr as well) when the uncertainty cost is taken into consideration, and this inferiority increases as the cost per unit of uncertainty increases.

8. Conclusion and future work

In this paper we considered dead-reckoning policies for updating the database location of moving objects, and the processing of range queries for motion database. When using a dead-reckoning policy, a moving object equipped with a Geographic Positioning System periodically sends an update of its database location and provides an uncertainty threshold th. The threshold indicates that the object will send another update when the deviation, namely the distance between the database location and the actual location, exceeds th.

Dead-reckoning policies imply that the DBMS answers a query about the location of an object m by: "the current location of m is (x, y) with a deviation of at most th". When making decisions based on such an answer, there is a cost in terms of the deviation of m from (x, y), and in terms of the uncertainty about its location. These costs should be balanced against the cost (in terms of wireless bandwidth, and update processing) of sending location update messages. We introduced a cost model that captures the tradeoffs between communication, uncertainty and deviation by assigning costs to an uncertainty unit, a deviation unit, and a communication unit. We explained that these costs should be determined by answering questions such as: how many messages is the system willing to utilize in order to reduce the deviation by one unit during a unit of time? Is a unit of uncertainty more important than a unit of deviation, or vice versa?

Then we introduced two dead-reckoning policies, adaptive dead-reckoning (adr), and disconnection detection dead-reckoning (dtdr). Both adjust the uncertainty threshold at each update to the current motion (or speed) pattern. This pattern is captured by the concept of the predicted deviation. The difference between the two policies is that dtdr uses a novel technique for disconnection detection in mobile computing, namely decreasing uncertainty threshold. Intuitively, the technique postulates that the probability of communication should

19

increase as the period of time since the last communication increases. Thus, the probability of the object being disconnected increases as the period of time since the last update increases. Dtdr demonstrates the use of this technique.

Then we reported on the development of a simulation testbed for evaluation of location update policies. We used it in order the compare the information cost of adr, dtdr, and speed dead-reckoning (sdr) in which the uncertainty threshold is arbitrary and fixed. The result of the comparison is that adr is superior to the other policies in the sense that it has a lower information cost. Actually, it may have an information cost which is six times lower than that of sdr. We quantified the disconnection detection cost as the difference between the cost of dtdr and that of adr. We also determined that when taking uncertainty into consideration, the information costs of adr and dtdr are lower than that of non-dead-reckoning policies which we developed previously.

Finally, an additional contribution of this paper is a probabilistic model and an algorithm for query processing in motion databases. In our model the location of the moving object is a random variable, and at any point in time the database location and the uncertainty are used to determine a density function for this variable. Then we developed an algorithm that processes range queries such as 'retrieve the moving objects that are currently inside a given polygon P'. The answer is a set of objects, each of which is associated with the probability that currently the object is inside P.

Now consider the following variant of the location update problem. In some cases MOD applications may not be interested in the *location* of moving objects at any point in time, but in their arrival time at the destination. Assume that the database arrival information is given by "The object is estimated to arrive at destination X at time t, with an uncertainty of U". In other words, t is the database estimated-arrival-time[9] (eat) and we assume that at any point in time before arrival at destination X, the moving object can compute the actual eat,[10] t'. The difference between t and t' is the deviation, and the uncertainty U denotes the bound on the deviation of the eat; the object will send an eat update message when the deviation reaches U. In this variant, the motion database update problem is to determine when a moving object should update its database estimated-arrival-time. The results that we developed in this paper for the location update problem carry over verbatim to the eat update problem. Actually, the location update problem is more general in the sense that the DBMS holds the estimated arrival time at any future location, not just the final destination.

Concerning future work, we believe that moving objects databases will become increasingly important, and we believe that DBMSs should be made a platform for developing moving-objects applications. For this purpose, as mentioned in the introduction, much remains to be done in terms of spatio-temporal query languages, support for rapidly changing real-time data, indexing, and imprecision.

Another issue is to extend the present work to handle uncertainty for moving objects that do not report their location; instead their location is sensed by possibly unreliable means. This is the case, for example, for enemy forces in a battlefield.

Appendix A: Comparison of the update policies by simulation

In this section we first discuss the simulation method and then the simulation results.

Simulation method

The objective of the simulations is to compare the information cost of sdr, adr and dtdr policies. As parameters to the simulation we use ten moving objects, each taking a two-hour trip. Each trip is represented by a speed curve, i.e., the actual-speed of a moving object as a function of time. In figures 2 and 3 at the end of the paper we show two typical speed curves.[11] The average distance traveled during a trip is 82 miles. For each speed curve, update policy, update cost C_1, and uncertainty unit cost C_2 we execute a simulation run. The run computes the information cost[12] (a single number) of the policy on the curve. (Observe that for the purpose of computing this information cost the actual route taken by the moving object is irrelevant; the deviation and the information cost at each point in time can be computed using only the speed curve.) Then, for each policy, we average the information cost over all the speed curves, and plot this average as a function of the update cost C_1.

Each simulation run is executed as follows. A speed-curve is a sequence S of actual speeds, one for each time unit. In our simulations a time unit is 10 s. Using S we simulate the moving object's computer working with a particular update policy. This is done as follows. For each time unit there is an uncertainty threshold th, as well as a database speed and an actual speed. The deviation at a particular point in time t is the difference between the integral of the actual-speed as a function of time, and the integral of the database-speed (the integrals are taken from the last update until t). Denote by T the sequence of deviations, one at each time unit. Denote by Q the sequence of uncertainty thresholds, one at each time unit. If the deviation at time t reaches the threshold, then we generate an update record

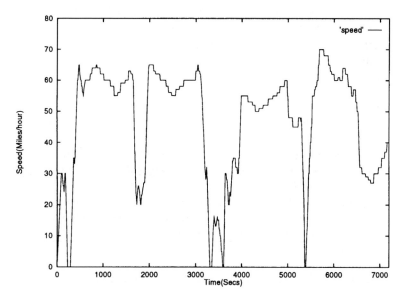

Figure 2. Speed curve for a two hour trip.

21

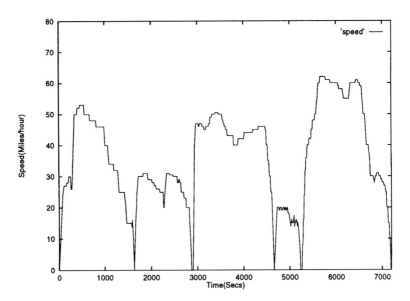

Figure 3. Speed curve for a two hour trip.

consisting of: the current time, the current location, the current speed, the next threshold (for adr and dtdr it is computed as explained in Section 5); the deviation at time t becomes zero. Denote by U the sequence of update records. Using T we compute the total cost of the deviation, denoted c_1, and using U we compute the total cost of the updates, c_2. Using Q we compute the total cost of the uncertainty, c_3. The information cost of the policy on the speed curve is $c_1 + c_2 + c_3$.

Simulation results

We compared the information cost for the adr, sdr, dtdr, and il (see [12]) policies. Figures 8–19 plot the results of the comparison. Each figure compares il, adr, dtdr and sdr(X), i.e., sdr with threshold X, where $X = 0.2, 1.0, 2.5$ or 7.5. Each figure uses a particular value of C_2. The C_2 values that we consider are 0.25, 0.75, and 3.00 (when $C_2 = 3$ it means that the cost of a unit of uncertainty is three times as high as the cost of a unit of deviation). When adr and dtdr are compared with sdr(X), the first (i.e., initial) uncertainty threshold of adr and dtdr is taken to be X; the following thresholds are determined dynamically, as explained in Section 5. The parameter X is irrelevant for the il policy because il is not a dead-reckoning policy. Each figure compares the information cost of the four policies. Each curve in figures 8–19 plots the information cost of a policy as a function of the update cost C_1.

For information about each cost component, figures 4–7 plot the information cost, the cost of updates, the cost of uncertainty, and the cost of deviation for the parameter setting $X = 2.5$ and $C_2 = 0.75$. Observe that since the number of messages used by sdr is independent of the

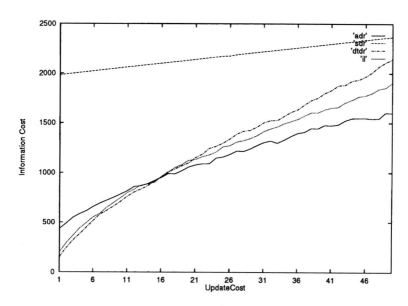

Figure 4. Average information cost for IL, ADR, SDR and DTDR with threshold = 2.5, UncerCost = 0.75.

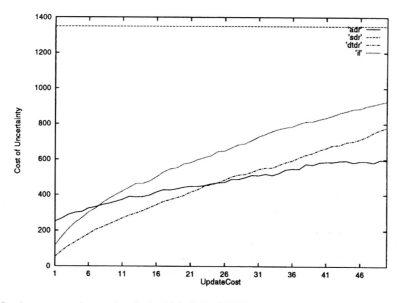

Figure 5. Average cost of uncertainty for IL, ADR, SDR and DTDR with threshold = 2.5, UncerCost = 0.75.

23

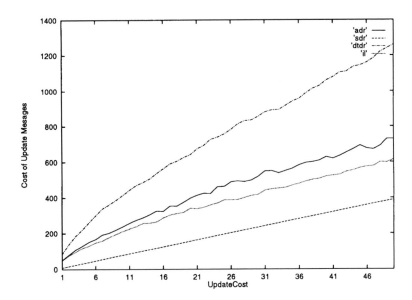

Figure 6. Average cost of update massages for IL, ADR, SDR and DTDR with threshold = 2.5, UncerCost = 0.75.

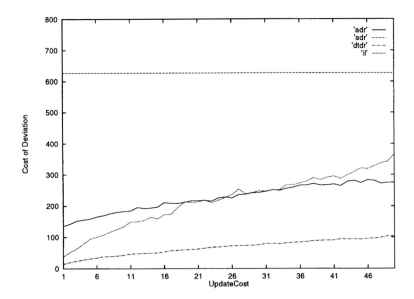

Figure 7. Average cost of deviation for IL, ADR, SDR and DTDR with threshold = 2.5, UncerCost = 0.75.

24

cost of a message, the uncertainty cost of sdr is independent of the cost of a message. The same holds for the deviation cost. On the other hand, the other policies adapt the number of messages to the message cost. Thus, they use less messages as the cost of a message increases, and consequently, for them, the deviation and uncertainty costs increase as the update cost increases. Observe that although the cost of a unit of uncertainty is lower than the cost of a unit of deviation (0.75 versus 1.00), uncertainty is the highest cost component for il (which is not true for the other policies) and it is higher than the uncertainty cost of the other policies.

Figures 8–19 are arranged such that each consecutive three give the information cost for a particular uncertainty threshold X, for the three values of C_2. Thus, for example, figures 8–10 give the information cost for $X = 2.5$. The basic conclusion from the simulations is that for almost all the experiments, the adr policy is superior to the other policies. The information cost of sdr is often several times higher than that of adr. One exception is the extreme threshold value $X = 7.5$, where the costs of uncertainty and deviation between the beginning of the trip and the next update dominate the total information cost of adr; this is an anomaly due to the unnecessarily high initial threshold. Concerning il, observe that since it cannot account for the cost of uncertainty, its total cost increases as the cost of uncertainty increases. Even for small values of C_2 it is not better than adr (except for the anomaluos threshold $X = 7.5$).

Observe that the information cost curves of the adr policy are almost identical for all the initial uncertainty thresholds. The same holds for the dtdr and il policies. This indicates that these policies exhibit a stable behavior which is independent of initial conditions.

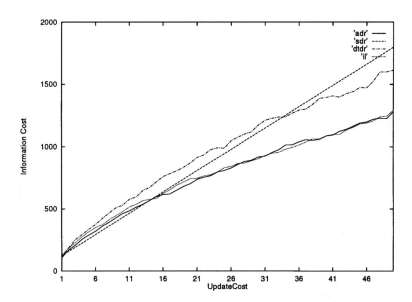

Figure 8. Average information cost for IL, ADR, SDR and DTDR with initial threshold = 0.2, UncerCost = 0.25.

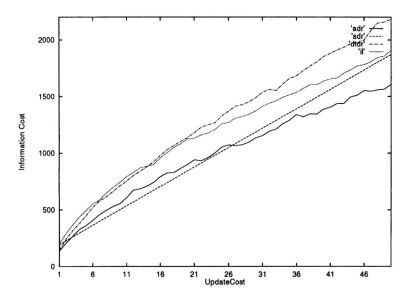

Figure 9. Average information cost for IL, ADR, SDR and DTDR with initial threshold = 0.2, UncerCost = 0.75.

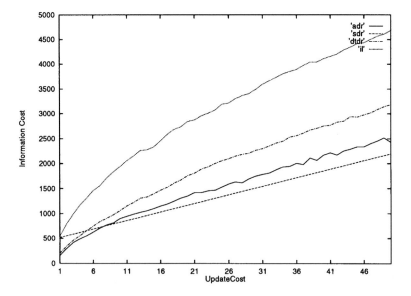

Figure 10. Average information cost for IL, ADR, SDR and DTDR with initial threshold = 0.2, UncerCost = 3.00.

26

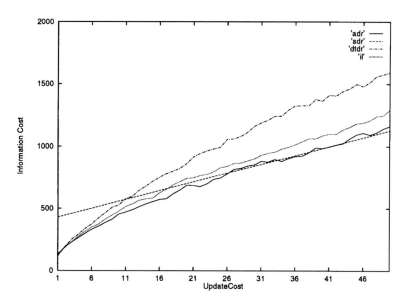

Figure 11. Average information cost for IL, ADR, SDR and DTDR with threshold = 1.0, UncerCost = 0.25.

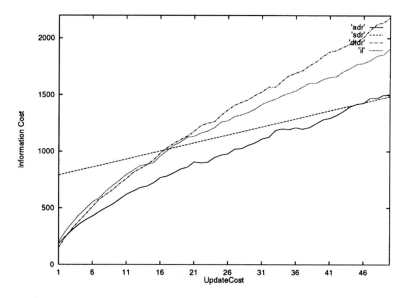

Figure 12. Average information cost for IL, ADR, SDR and DTDR with threshold = 1.0, UncerCost = 0.75.

27

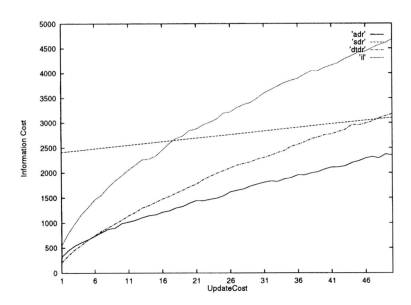

Figure 13.   Average information cost for IL, ADR, SDR and DTDR with threshold = 1.0, UncerCost = 3.00.

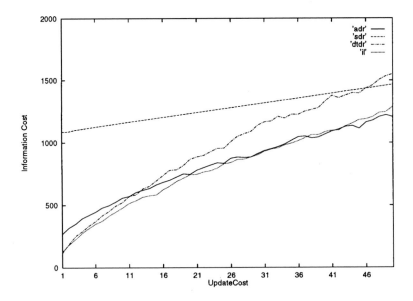

Figure 14.   Average information cost for IL, ADR, SDR and DTDR with threshold = 2.5, UncerCost = 0.25.

28

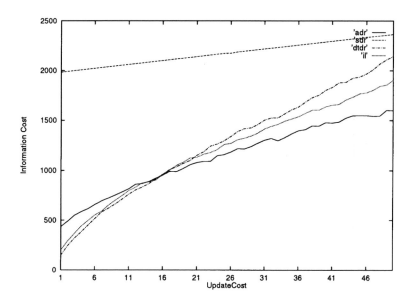

Figure 15.   Average information cost for IL, ADR, SDR and DTDR with threshold = 2.5, UncerCost = 0.75.

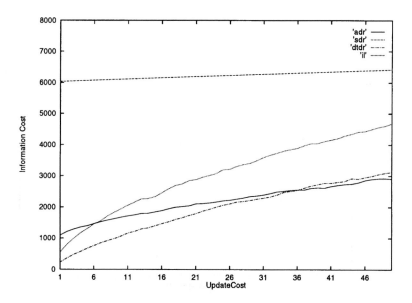

Figure 16.   Average information cost for IL, ADR, SDR and DTDR with threshold = 2.5, UncerCost = 3.00.

29

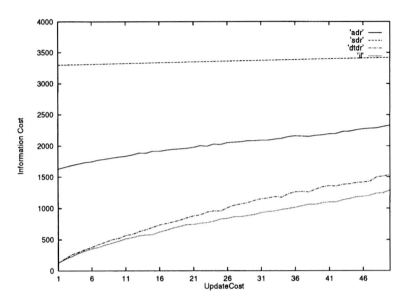

Figure 17.   Average information cost for IL, ADR, SDR and DTDR with threshold = 7.5, UncerCost = 0.25.

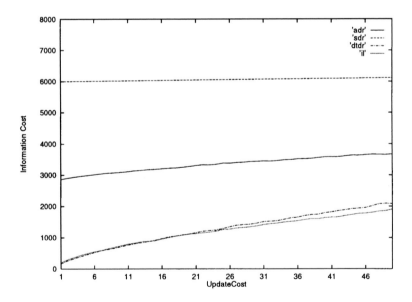

Figure 18.   Average information cost for IL, ADR, SDR and DTDR with threshold = 7.5, UncerCost = 0.75.

30

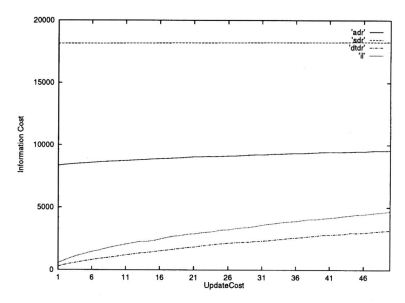

Figure 19.   Average information cost for IL, ADR, SDR and DTDR with threshold = 7.5, UncerCost = 3.00.

Appendix B: Demonstration of the adr and dtdr update policies

Demonstration of adr by an example.   At the beginning of the trip the moving object *m* sends a location update message giving the route, its location on the route, *L.speed* = 0.2 miles/min, and *L.uncertainty* = 0.5 miles. Suppose that after 4 min the deviation reaches the threshold 0.2. At that time *m* sends a location update message containing its current location on the route, its current speed, and a new value for *L.uncertainty*. Suppose that the integral of the deviation from the beginning of the trip (time 0) to time 4 (minutes) is 1, and $C_1 = 8$, and $C_2 = 1$. Then the new value for *L.uncertainty* is 0.82. Suppose that after 10 more minutes the deviation reaches this new threshold, and at that time the integral of the deviation from time 4 to time 14 is 1.5. Then the new value for *L.uncertainty* is 0.4.

Demonstration of dtdr by an example.   At the beginning of the trip, the moving object *m* sends a location update message giving the route, its location on the route, *L.speed* = 0.2 miles/min, and *L.uncertainty* = 0.5 miles. A time unit is 1 min. Then for the first minute the threshold is 0.5, and for the second minute the threshold changes to 0.25 miles. Suppose that during the second minute of the trip, the deviation reaches the current threshold 0.25. At that time *m* sends a location update message containing its current location on the route, its current speed, and a new initial value for *L.uncertainty*. That value is computed as follows. Suppose that the integral of the deviation from the beginning of the trip (time 0) to time 2 is 0.5, and $C_1 = 8$, $C_2 = 1$. Then the slope of deviation estimator is $a = \frac{2 \times 0.5}{2^2} = 0.25$, and the new value for *L.uncertainty* is 2.226 (it is computed by a numerical solution to the Eq. (9)). Suppose that after 5 more minutes, the deviation reaches the current threshold

31

which is $\frac{2.226}{5} = 0.445$, and at that time the integral of the deviation from time 2 to time 7 is 1.2. Then the slope of deviation estimator is $a = \frac{2 \times 1.2}{5^2} = 0.096$, and the new initial *L.uncertainty* is 1.992.

Notes

1. Our simulation experiments show that, even when the speed fluctuates sharply, this temporal technique reduces the number of updates to 15% of the number used by the traditional, nontemporal method in which the database simply stores the latest known location for each object; this saves 85% of the location-updates overhead.
2. We use the term *speed* dead-reckoning to contrast it with the plain dead-reckoning (pdr) policy in which the database location is fixed until it is explicitly updated by the moving object; namely, pdr does not use dynamic attributes.
3. Another possibility for representing future locations is a sequence of speeds, i.e., the object will move at speed v_1 until time t_1, at speed v_2 until time t_2, etc. Such a future plan is typical of, for example, a vehicle that expects various traffic conditions; or a package that first travels by truck, then by plane, then waits (speed 0) for another truck loading, etc.
4. Let us observe that the proposed method of optimizing the new threshold K is not unique. We have devised other methods which are omitted from this extended abstract. A performance comparison among these methods is the subject of future work.
5. Sdr can also use another speed, for example, the average speed since the last update, or the average speed since the beginning of the trip, or a speed that is predicted based on knowledge of the terrain. This comment holds for the other policies discussed in this section.
6. When comparing the adr and sdr policies by simulation (see the Appendix), both policies use as their first threshold the same value, and the same holds for the initial speed.
7. More powerful, nonlinear approximation functions have been considered, but their discussion is omitted from this extended abstract.
8. In this sense, we are using a simple linear regression, but instead of the common least squares method we employ an equal sums method that is more appropriate for our cost function.
9. Equivalent to the database location.
10. Equivalent to the actual location.
11. The speed curves were generated using an approximation of the speed during a trip on highways and local streets in the Chicago area, for various times of day (e.g., rush hour, night, etc.).
12. Remember, the *information cost* of an update policy on a given speed-curve is computed by using Eq. (3) for every time interval between two consecutive update points.

References

1. S. Abiteboul, R. Hull, and V. Vianu, Foundations of Databases, Addison Wesley, 1995.
2. R. Alonso and H.F. Korth, "Database system issues in nomadic computing," in Proceedings of the 1993 ACM SIGMOD International Conference on Management of Data, Washington DC, May 1993.
3. M. Abadi and Z. Manna, "Temporal logic programming," Journal of Symbolic Computation, vol. 8, 1989.
4. B.R. Badrinath, T. Imielinski, and A. Virmani, "Locating strategies for personal communication networks," Workshop on Networking for Personal Communications Applications, IEEE GLOBECOM, December 1992.
5. M. Baudinet, M. Niezette, and P. Wolper, "On the representation of infinite data and queries," ACM Symposium on Principles of Database Systems, May 1991.
6. A. Brodsky, V.E. Segal, J. Chen, and R.A. Exarkhopoulo, "The CCUBE constraint object-oriented database system," manuscript, 1997.
7. J. Chomicki and T. Imielinski, "Temporal deductive databases and infinite objects," ACM Symposium on Principles of Database Systems, March 1988.
8. W. Feller, An Introduction to Probability Theory, John Wiley and Sons, 1966.

9. M. Goodchild and S. Gopal (Eds.), Accuracy of Spatial Databases, Taylor and Francis, 1989.

10. J.S.M. Ho and I.F. Akyildiz, "Local anchor scheme for reducing location tracking costs in PCN," first ACM International Conference on Mobile Computing and Networking (MOBICOM'95), Berkeley, California, November 1995.

11. P. Sistla, O. Wolfson, S. Chamberlain, and S. Dao, "Modeling and querying moving objects," in Proceedings of the Thirteenth International Conference on Data Engineering (ICDE13), Birmingham, UK, April 1997.

12. O. Wolfson, S. Chamberlain, S. Dao, L. Jiang, and G. Mendez, "Cost and imprecision in modeling the position of moving objects," in Proceedings of the Fourteenth International Conference on Data Engineering (ICDE14), 1998.

13. S. Grumbach, P. Rigaux, M. Scholl, and L. Segoufin, "DEDALE: A spatial constraint database," manuscript, 1997.

14. T. Imielinski and H. Korth, Mobile Computing, Kluwer Academic Publishers, 1996.

15. R. Jain, Y.-B. Lin, C. Lo, and S. Mohan, "A caching strategy to reduce network impacts of PCS," IEEE Journal on Selected Areas in Communications, vol. 12, October 1994.

16. P. Kanellakis, "Constraint programming and database languages," ACM Symposium on Principles of Database Systems, May 1995.

17. D. Lazoff, B. Stephens and Y. Yesha, "Optimal location of broadcast sources in unreliable tree networks," IEEE International Conference on Computers and Communications Networks, Rockville, Maryland, 1996.

18. A. Motro, "Management of uncertainty in database systems," in Modern Database Systems, Won Kim (Ed.), Addison Wesley, 1995.

19. OmniTRACS, "Communicating without limits," http://www.qualcomm.com/ProdTech/Omni/prodtech/omnisys.html.

20. J. Paradaens, J. van den Bussche, and D. Van Gucht, "Towards a theory of spatial database queries," ACM Symposium on Principles of Database Systems, May 1994.

21. R. Snodgrass and I. Ahn, The temporal databases, IEEE Computer, Sept. 1986.

22. S. Chamberlain, "Model-based battle command: A paradigm whose time has come," 1995 Symposium on C2 Research & Technology, NDU, June 1995.

23. S. Chamberlain, "Automated information distribution in bandwidth-constrained environments," MILCOM-94 conference, 1994.

24. S.D. Silvey, Statistical Inference, Chapman and Hall, 1975.

25. N. Shivakumar, J. Jannink, and J. Widom, "Per-user profile replication in mobile environments: algorithms, analysis, and simulation results," ACM/Baltzer Journal on Special Topics in Mobile Networks and Applications, special issue on Data Management, 1997.

26. Y. Yesha, K. Humenik, P. Matthews, and B. Stephens, "Minimizing message complexity in partially replicated databases on hypercube networks," accepted for publication in Networks, in press.

Distributed and Parallel Databases, 7, 289–317 (1999)
© 1999 Kluwer Academic Publishers. Manufactured in The Netherlands.

Mobility and Extensibility in the StratOSphere Framework*

DANIEL WU   danielw@cs.ucsb.edu
DIVYAKANT AGRAWAL   agrawal@cs.ucsb.edu
AMR EL ABBADI   amr@cs.ucsb.edu
Department of Computer Science, University of California, Santa Barbara

Received October 8, 1998; Accepted February 26, 1999

Abstract. We describe the design and implementation of our **StratOSphere** project, a framework which unifies distributed objects and mobile code applications. We begin by first examining different mobile code paradigms that distribute processing of code and data resource components across a network. After analyzing these paradigms, and presenting a lattice of functionality, we then develop a layered architecture for **StratOSphere**, incorporating higher levels of mobility and interoperability at each successive layer. In our design, we provide an object model that permits objects to migrate to different sites, select among different method implementations, and provide new methods and behavior. We describe how we build new semantics in each software layer, and present sample objects developed for the Alexandria Digital Library Project at UC Santa Barbara, which as been building an information retrieval system for geographically-referenced information and datasets. We have designed using **StratOSphere** a repository that stores its holdings. The library's map, image and geographical data are viewed as a collection of objects with extensible operations. **StratOSphere**.

1. Introduction

The advent of distributed computing has had a major influence in the computing industry in recent years, witnessed by the growth of mobile computers and networked computing systems. The desire to share resources, to parcel out computing tasks among several different hosts, and to place applications on machines most suitable to their needs has led to distributed programming systems such as CORBA [35] and DCOM [5] that predominate in the marketplace. Despite competing standards, both systems have very similar designs. They each define a distributed object model that achieves interoperability through strict separation between interface and implementation. A CORBA or DCOM object registers its IDL (Interface Definition Language) interface into an Interface Repository. The IDL interface specifies method signatures for each object, describing in part its behavior and semantics. Applications then obtain references to each distributed object, and invoke operations based upon these method signatures.

The benefits of this approach are evident: These systems provide location transparency, so that applications residing on different hosts may invoke one another; the object model

*Work supported by research grants from NSF/ARPA/NASA IRI9411330 and NSF CDA-9421978.

in each system promotes encapsulation of program code so as to facilitate management of remote objects; the separation between IDL interface and code implementation permits implementations in different languages through different language bindings.[1] The drawback, however, is that mobile code applications are not well supported. As of yet, there exist no standardized services in either CORBA or DCOM to migrate objects and threads for distributing processing, no facilities for customizing or providing new object behavior at run-time, no provisions for agent processing or autonomous computing. Although these features can certainly be grafted on top of CORBA and DCOM systems, doing so would add an even more cumbersome layer of services onto—especially in the case of CORBA—an already bulky software package.

In seeking to support mobile code applications, we take a different approach: We first examine the requirements of mobile code systems, and develop a corresponding software system with higher-level abstractions of *code*, *data*, and *migration approaches*. To achieve this result, we describe the design of the **StratOSphere** architecture, which supports mobile applications written in Java. Every entity in **StratOSphere** is modeled as a Java object. In the Alexandria Digital Library (ADL) project at UC Santa Barbara [36], for example, the library's collection of high resolution aerial photographs, satellite images, and topographical maps would each be modeled as an object, replete with methods to filter, layer, and display the object instance. That is, a library document is no longer a passive entity, but an object that can be instantiated and processed through method operations. An image, for example, belonging to a `LandsatImage` class has as relevant methods: `zoomIn()`, `getThumbnail()`, `queryElevation()`, `print()`.

The library's current holdings consist of over 700,000 catalogue items, with some images exceeding a gigabyte of storage. The design of a digital library for this collection requires that the vast amounts of data be distributed across different hosts. Furthermore, not only must the data be distributed, but the processing of the data must also be distributed. To address these issues, we have developed a Data Store for Alexandria objects using **StratOSphere**, that permits both object instances and object methods to be migrated to different sites for remote execution. In our design, we provide objects with the capability of obtaining new functionality by dynamically locating and processing new executable code.

The paper is organized as follows: Section 2 decribes the management of distributed objects in an information system; Section 3 describes related work on distributed programming; Section 4 describes the requirements of distributed object management; Section 5 describes the design and implementation of **StratOSphere**; Section 6 describes the data model of the Alexandria Digital Library Project; and Section 7 concludes the paper.

2. Distributed objects in an information system

Within a distributed object-based digital library, data and resources can be scattered and co-located to several hosts. When processing of these resources take place, there are two basic alternatives: either transfer all the data items to a centralized access point for processing, or distribute the processing by visiting each host and executing a portion of the execution at the remote host. There are trade-offs to either approach depending on the size of the data items, and the amount of processing capability at each site. In our distributed system

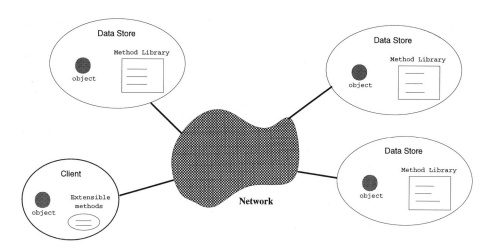

Figure 1. Distributed objects propagated across data store servers.

(figure 1), where code and data mobility are primary considerations, we provide support for both forms of access and processing. We begin by discussing the basic requirements for true object distribution and extensibility in a digital library.

2.1. Object migration

Consider a situation where a Client application creates a large LandsatImage object at a Data Store Server and later that Server becomes so heavily loaded that it can no longer provide adequate performance. Rather than continue processing at the Server, the Client would like to download the object for local processing, or migrate that object to another less heavily-loaded Server. Alternately, consider a distributed network in which services and resources are available at remote sites. Again, the need to migrate the object becomes evident. Current distributed technologies support object migration by transferring the state of an object between two different hosts. CORBA, for example, permits applications to transfer an object's state, but not its actual implementation. This restriction serves several purposes; it provides a framework for language-independent and host-independent implementations. The drawback, however, is that while the state of an object is migrated, its processing is not. An application still accesses the object's methods through a proxy; the execution is performed on the host where the operating code is stored. Thus, while information pertaining to the object may be distributed, its processing is still restricted to the target platform where the implementation code was registered.

To fully support object migration services, an object's methods, as well as its state must be shipped together as a single programming unit, so that the receiver may perform processing upon that object at the remote site. Providing this capability, though, forces many requirements upon the system. The receiver must be able to run the sender's object code, thus fixing the binary and run-time executable for the object implementation across separate

37

machines. To achieve platform independence, we have chosen Java [2] as our implementation language because both its byte-code and Java Virtual Machine are designed to be machine-independent. Programming in Java adds a virtual programming layer upon the native operating environment. In addition, the Java programming architecture is standardized so that Java objects may be serialized into portable byte-code using JDK 1.1's Object Serializer [30]. The choice of Java thus addresses many of the platform compatibility issues raised in implementing a distributed system.

2.2. *Object extensibility*

Object roles and behavior [29] in a digital library evolve over time, and may need to be individually tailored to specific user requirements. Hence, a distributed digital library must provide some means of dynamically extending an object's functionality. For example, the `LandsatImage` object supplies a `locateVegetation()` method to segment and outline the areas of vegetation growth. Suppose, though, that a particular Client would like a `locateLivestock()` method to identify bovine grazing areas. Rather than create a new object class with such functionality and register it into the Data Store repository, we need to provide the Client with the capability of extending an object's methods. The Client can write its own `locateLivestock()` method and ship it to the object, thus extending the object's available methods. This is an example of a Client providing its own private object extensions. The Data Store object is temporarily extended to a new object to satisfy a user's particular need. A trusted user (a librarian user), however, could introduce a new object into the Data Store by defining a new subclass in the Data Store repository; this would then result in a permanent and global change, visible to all Clients.

2.3. *Dynamic object access*

Another limitation found in current tools is the requirement of a static proxy to access a distributed object. In systems such as CORBA, an object is first created and registered into an interface repository and implementation repository. A Client program queries the interface repository to obtain a proxy to the remote object. Execution takes place by calling methods through the proxy. This manner of processing follows a rigid imperative style of programming, in which access is provided only after an object proxy has been located. For our system, we offer a more dynamic discovery-based form of access in which a Client can query Data Store repository records to determine the structure, format, and meta-data of Data Store objects. Upon discovering an object type of interest, the Client can instantiate the object directly without reference through a proxy. The Data Store repository can thus be viewed as an extended database in which data, object state, and object methods are stored in a distributed manner.

2.4. *Dynamic views of distributed objects*

Just as each Data Store holds a repository of method programs, a digital library database should also let the Client build its own private repository to reflect its view of the relations among the Data Store items. By storing object instances and object methods within a migrating store, the Client can process its contents by visiting selected Data Store sites to collect new data and executable resources. The implementation of this view, however, requires additional programming facilities. For this reason, we introduce a programming model for Data Store applications that utilizes the Data Store's migration services. In this model, an application constructs a view of a mobile Data Store object. The program instructions direct the object to visit Data Store sites in the network, executing and processing data along each site. The goal of this model is to distribute resources and processing in the network so that an object completes a computational task by visiting a set of Data Store servers.

3. Related work

Work in distributed objects and interoperability has had a long history, beginning with attempts to extend programming to remote hosts in order to provide greater collaboration and sharing of resources. These efforts can be classified into three basic categories: distributed programming libraries and packages, distributed operating systems [13], and distributed programming languages. These categories reflect attempts to extend a programming environment by adding a distribution layer to an existing language, by providing distributed access as an operating system primitive, and by developing a language with fully distributed scope and semantics. These designs each have their various advantages and disadvantages, depending on the amount of programming effort required upon the application developer, the amount of specialized infrastructure needed to provide an interoperability layer, and the ease of extending an existing programming system.

3.1. Distributed programming systems

Distributed programming systems provide the simplest form of distributed programming support, by adding a distribution layer on top of a language system. They usually provide remote stubs at each end of hosts that wish to communicate and perform remote invocations. Systems such as RPC [4], DCE [27], CORBA [35], DCOM [5], RMI [44], and HORB [33] each generate proxies to provide distributed access to remote resources. While RMI and HORB are Java-specific, RPC, DCE, CORBA, and DCOM provide language bindings for a particular language implementation. In doing so, they are able to unify different language systems to build common applications. Their main advantage is that they require relatively little effort to incorporate into an existing software system; their disadvantage is that the limited form of interoperability and object migration that they support, although CORBA, DCE, and DCOM seek to ameliorate this by providing a rich set of standard services for creating, locating, registering, and storing an object.[2]

3.2. Distributed operating systems

The operating system approach addresses a different aspect of distribution by migrating processes for load-balancing, fault-tolerance, and resilience. Systems such as Sprite [11], V [9], and Locus [28] provide built-in operating system support to freeze the run-time computation of a process, migrate the process to a remote site, and unfreeze the process. Each migration entails leaving a proxy behind to forward I/O and communications to the new site. By providing fine-grained mobility, the location of processors, memory, and file systems can be made fully transparent to an application program. Though providing an extremely powerful and elaborate design, users would have to upgrade and install an entirely new operating system to their existing computer system—a prohibitive modification in current enterprises.

3.3. Distributed languages

Among distributed languages, there are languages designed with fully distributed semantics such as Obliq [6], Emerald [22], and Distributed Oz [41], and adaptive mobile code languages such as TACOMA [21], Telescript [43], Sumatra [1], Agent TCL [15], Odyssey [42], Voyager [26], and the Liquid Software project at University of Arizona [18]. In the former category are languages whose semantics were designed to support distributed scope and access, provide a shared memory abstraction, and extend ordinary language operations to manage replicated objects and data. The latter category comprise agent languages and mobile code frameworks which provide object mobility within a distributed environment.

The Liquid Software project, in particular, introduces the notion of dynamically moving functionality within a network to enable adaptive computing. Consisting of an array of retargetable compilers, customizable client/server interfaces, and operating system support library, the Liquid Software infrastructure can be used to develop software agents for browsing, searching, and retrieval. In **StratOSphere**, we take a very similar approach by developing a distributed object system that provides network transparency for the location of data and the location of its execution; in **StratOSphere**, however, we provide an *object repository* that resides at each site where processing takes place. We show how remote object repositories consisting of instances and methods permit objects to take on dynamic functionality by acquiring new methods and behavior at these remote sites.

4. Mobile processing and mobile objects requirements

Before describing the architecture of **StratOSphere**, we first examine the forms of mobility provided by various distributed and network programming packages. We analyze different execution scenarios to determine what form of mobility is required of the distributed system.

4.1. Mobile code paradigms

In an earlier work, Carzaniga, Picco, and Vigna [7, 10] provide an elegant description of several mobile code design paradigms for distributed applications. These are classified as *Client/Server* (CS), *Remote Evaluation* (REV), *Code on Demand* (COD), and *Mobile Agent* (MA) paradigms. By decomposing distributed applications into *resource components* (code and data), *computation components* (thread of execution), *interactions* (event and information passing between two or more components), *sites* (location where processing

Table 1. Mobility and remote execution paradigms.

Paradigm	Before migration		After migration	
	S_A	S_B	S_A	S_B
CS	A, D_A, C_A	B, D_B, C_B	A, D_A, C_A	B, D_B, C_B, D_A
REV	A, D_A, C_A	B, D_B, C_B	A, D_A, C_A	B, D_B, C_B, D_A, C_A
COD	A, D_A, C_A	B, D_B, C_B	A, D_A, C_A, C_B, D_B	B, D_B, C_B
MA	A, D_A, C_A	$-D_B$, C_B	$-D_A$, C_A	A, D_B, C_B, D_A, C_A

takes place), distributed execution can be modeled as primitives operating in one of the above mobile code scenarios.

We briefly describe each mobile code presentation as follows: Computation components A and B reside at sites S_A and S_B, and component A initiates some interaction with component B. Code and data resources from components A and B, (C_A, D_A, C_B, D_B) are presented below before and after each migration in Table 1.

In the CS paradigm, the code, data, and execution remain fixed at the server site S_B. This is the usual RPC style of programming in which a client requests a service of a server by specifying some data resource D_A. The program code C_B and remaining data resources D_B to perform this service are resident with the server. An additional interaction from B to A returns the result of the execution.

In the REV scenario, the code to perform the execution is stored at S_A. Component A ships both code and data to site S_B where it gets processed at S_B. A final interaction returns the result from B to A. Although mobile code scenarios CS and REV are very similar in nature, they differ in one key aspect: the ability to transfer processing from one site to another by transferring the executable code. In languages such as LISP, Scheme, or any of a variety of scripting languages, REV can be seen as a natural extension of CS, since data (represented as lists, strings, or tuples) can just as easily be interpreted as code. That is, C_A can be represented as D_A. As long as a program interpreter is present at each server site, remote evaluation and execution are readily available. For procedural languages which distinguishes between programming instructions and data structures, however, the run-time requirements of REV differ markedly from those of CS. The additional requirements of shipping code from one site to another affects the environment's architecture and implementation semantics.

Scenario COD is an inversion of REV. Instead of initiator A sending code and data off to B, component A requests code and data from B, and executes it locally. An example of this form of code mobility is the applet service in the Java programming language, in which HTTP browsers download Java code from remote sites for local execution. An additional service known as *servlet* programming [8] performs the opposite service, pushing Java code from a local client to a remote server for remote execution, thus falling under the REV category. While COD and REV are very similar in design, the performance differences between the two often lead to one scenario being preferred over the other. In Web computing, the applet model COD proliferates, as the bulk of processing requires that computations be offloaded to client sites. However, in more complex interactions, servlets and CGI [16] programming perform execution at the server-side, where relevant data and resources are stored.

Finally, MA, is the most dynamic and autonomous of the above paradigms. In the three previous scenarios, code and data components can be transferred from one site to another to re-direct the location of processing, but the computation component (the executing process) remains fixed to its original site. In CS, REV, and COD, initiator A begins an interaction and obtains the result in a separate thread from that of the receiving component B. In MA, however, not only can resource components be transferred, but so can the entire computational component. Indeed, from Table 1, we observe that MA is an extension of REV, in which not just D_A and C_A are migrated, but also the computational component A itself. To perform this service, the mobile agent A suspends its execution at site S_A, preserves its state into an *execution unit* (EU), and transfers the EU to site S_B where processing resumes from A's recent state.

These four mobile code paradigms are among the types of mobility we support in **StratOSphere**. We introduce another mobile code paradigm called RCE (Remote Code Execution), which is a union of REV and COD mobility; a RCE facility is able to transfer code C_A and data C_A to a remote site for remote execution, as well as pull remote code C_B and data D_B for local execution. They can be organized into the following lattice (figure 2)

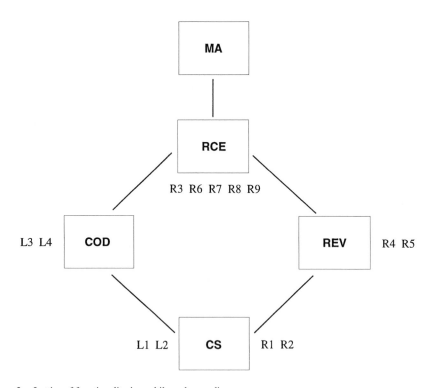

Figure 2. Lattice of functionality in mobile code paradigms.

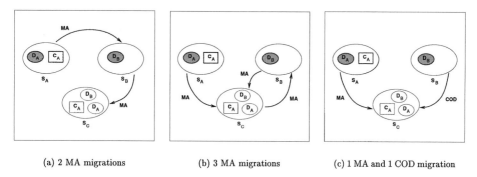

(a) 2 MA migrations   (b) 3 MA migrations   (c) 1 MA and 1 COD migration

Figure 3.   MA and MA/COD migrations.

to evaluate the mobile computing capability of each scenario. The least functional mobile code paradigm is CS, in which both code and computational components are fixed, though data components are migratory. In REV, COD, and RCE scenarios, both code components and data components are mobile but the computational components remain fixed. All three components are mobile in the most functional mobile code paradigm of all, MA. In principle, it may appear that MA is the only mobile code paradigm required, as it provides migration of all computational components. Indeed, mobile agent technologies advance this sup-position. By considering the costs of extra copying, however, we examine the overhead encumbered by MA.

Suppose code and data components C_A and D_A are stored at site S_A while data component D_B is stored at site S_B, but execution is to take place at site S_C (figure 3); assume further that the initiating process resides at site S_A. Adopting a sole MA approach, the execution unit may load in C_A and D_A, then visit site S_B to load in D_B (figure 3(a)), before finally arriving at S_C to execute C_A with D_A and D_B. The drawback to this approach is that components C_A and D_A are migrated twice: from S_A to S_B, then from S_B to S_C; if the data item D_A is very large then the performance penalty that MA incurs may be prohibitive.

To alleviate the costs of transferring large data components, MA may be employed in a different manner. After loading in C_A and D_A, the execution unit first visits S_C and deposits C_A and D_A in a repository at S_C; next the execution unit migrates to S_B where it collects D_B and returns to S_C where execution can proceed with code and data components are in place. This scenario (figure 3(b)) requires three migrations of MA. While the overhead of unnecessarily transferring large data objects has been eliminated, this approach also incurs a performance penalty, albeit less evident than the preceding scenario. In depositing the components C_A and D_A at S_C, then migrating the EU to S_B and back to fetch D_B, the EU must initially unload components C_A and D_A into a repository at S_C and then reload those same components before execution can proceed. We note further the two MA migrations to S_B and back are for the purpose of fetching components, a feature for which the COD mechanism was designed. By replacing these two MA migrations with a single COD transfer (figure 3(c)), the execution unit can thus avoid having to insert and subsequently retrieve C_A and D_A from storage at S_A. Thus, though MA provides a highly desirable form of mobility, the remaining forms of mobility cannot be discarded. In designing **StratOSphere**, we provide a general migration facility that supports all the above forms of mobility, in order to adapt to a number of different execution environments.

4.2. Execution scenarios

The mobile code paradigms presented earlier illustrated the effects of mobility by fixing the locations of processing and migrating code and date resources. We now take the alternate approach of fixing the locations of code and data placement, and migrating the processing. As Semeczko and Su [34] have pointed out, there are 13 possible data and execution location scenarios for data and executable programs stored at local and remote hosts. In these scenarios, both the data and executable can be migrated to different hosts for remote processing. Table 2 lists the scenarios for *local execution*, while Table 3 lists the scenarios for *remote execution*. In these tables, *L* represents the local host, and *R* the remote host, while *E* represents the host where execution is to take place.

The four local execution scenarios (L1–L4) depict those scenarios where code and data are to be pulled to the local site for local execution. They require a COD facility to satisfy their mobility requirements. Of these scenarios, L1 and L2 require only CS for data passing, which is readily achieved through client/server distributed computing packages such as CORBA, DCE, and RPC.

Table 2. Local execution scenarios.

	Location of code	Location of data	Mobile paradigm
L1	L	L	CS
L2	L	R	CS
L3	R	L	COD
L4	R	R	COD

Table 3. Remote execution scenarios.

	Location of code	Location of data	Mobile paradigm
R1	E	E	CS
R2	E	L	CS
R3	E	R	RCE
R4	L	E	REV
R5	L	L	REV
R6	L	R	RCE
R7	R	E	RCE
R8	R	L	RCE
R9	R	R	RCE

The remote execution scenarios, however, require a greater degree of mobility and functionality. Since the component initiating the processing does not reside at the execution site, a COD mechanism does not suffice. For these remote executions, four of the cases (R1, R2, R4, R5) can be implemented through the use of REV frameworks, of which two (R1 and R2) require only CS. For the remaining five cases, we proceed higher up the mobility ordering to obtain RCE functionality. In each of the cases (R3, R6, R7, R8, R9), a RCE facility can be employed to visit first the execution site, and from there pull any remaining code and data from a remote site to the execution site for processing.

Altogether, these 13 scenarios comprise the possible combinations of executions in a distributed environment, where code and data can both be migrated for mobile code applications. Supporting such functionality, however, is only one aspect of a distributed object system; mobility serves distribute processing throughout a network, but we need further facilities to manage the creation and execution of mobile objects.

4.3.   External object methods

In **StratOSphere**, an object's methods may be stored separately from the object instance. We view this as a fundamental requirement in permitting objects to adapt to changing conditions in a fluid distributed environment where changes in operating conditions are inevitable. By permitting an object's methods to be substituted at run-time, an object can modify and extend its services without having to recompile.

At first glance, this treatment may appear surprising to adherents of object-oriented programming, as an object is traditionally understood to store both data (object state), and executable code (object method) as a single programming unit; thus, decoupling an object's state and methods seems contrary to an object-based perspective. On the one hand, the demands of mobile and distributed computing environments promote the separation of large object data and object services, while on the other hand, the object-oriented approach requires their co-location. To reconcile these two views, we model each object as a loosely-coupled object. That is, an object consists of *internal methods* that are bound to an object, as well as *external methods* that operate on the object state, but are stored externally. In programming terms, an internal method is a method that resides in a class, while an external method resides in a separate executable program as a Java `.class` file. Internal methods are those methods that provide consistent modification of object state, while external methods provide extended services. At run-time, a Java object's internal method may be invoked by means of the Reflections package [38], while an external method can run upon an object by using a Java class loading mechanism [14]. We explore these concepts in greater detail below.

4.3.1. Run-time object adaptation.   Beyond resource and computation migration, the ability to adapt a running object to changing needs and requirements remains an essential aspect of an interoperable object system. For this we require the ability to extend the services of an object. For example, consider a scenario in which a client wishes to process an aerial image of a region to determine areas of vegetation growth. The **StratOSphere** object in the system, called `Image`, contains methods to `read` in an image, `resize`, and `display`

45

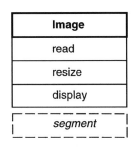

Image
read
resize
display

SegmentableImage
read
resize
display
segment

Image
read
resize
display
segment

(a) A StratOSphere class   (b) Extend class's functionality   (c) Dynamically extends
                                  by subclassing from Image   the method of a class

Figure 4.   Extending a method in **StratOSphere**.

the image. The client, furthermore, would like to segment the aerial image using image processing routines to highlight the areas of vegetation growth.

In traditional object systems, the client would usually subclass from the Image class in order to extend its functionality by providing a segment method. Creating a new subclass, however, is a compile-time activity; the client declares a new class, SegmentableIm-age, that inherits from the base class, adding and redefining new methods and instance variables as appropriate. This technique works well in object-oriented languages, because the programmer can freely extend classes and introduce new types and features into the programming environment. This capability, however, cannot be sustained in distributed environments, where each user is limited in the ability to add or modify system classes. Rather than provide a new subclass at compile-time, the **StratOSphere** system lets users supply a new method to an object at run-time (figure 4).

In many environments, security restrictions forbid a client from arbitrarily introducing a new type into the system. For applications, however, this inability to modify or extend the services of an object, poses too rigid and unwieldly a restriction. As time progresses, the role of an object evolves and adapts [29], and a software system must accommodate these changes by providing the necessary user enhancements and modifications. For this reason, we too must provide a means of dynamically extending an object's services in our **StratOSphere** system. In doing so, we must ensure that redefining or adding a new method does not lead to an inconsistent type system for our objects. As we shall see, the means by which we ensure consistency is through run-time assertions of methods and class invariants.

4.3.2. Repository re-implementation. Taking this method one step further, we not only permit an applications to extend its object behavior, we provide repositories with this capability as well. Specifically, we let object repositories provide different implementations of external methods.

Suppose after some time, a SegmentableImage class gets introduced into the **StratO-Sphere** environment (figure 5(a)). A client application may still not be satisfied with the given segment method. Once again, the client application can provide its own

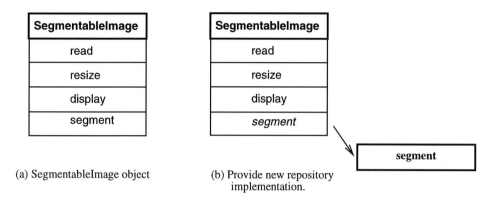

(a) SegmentableImage object

(b) Provide new repository
    implementation.

Figure 5. Repository redefines method with new implementation.

implementation by dynamically extending the class and redefining the segment operation, or better yet, the client can query a global directory to locate an already existing
repository method in which to redefine the operation (figure 5(b)). The signature of the
relevant method can be ascertained from the directory, and must be compatible with the
actual formal arguments, lest a run-time **StratOSphere** exception is thrown.

In this case, we let the repository rather than the object reimplement a method in order
to obtain new functionality or take advantage of a particular host repository's resources.
By letting both applications and repositories specialize their services, the **StratOSphere**
environment provides convenient adaptation for both source program and distributed environment, presenting a customizable run-time environment to adjust to changing conditions. We shall see how we obtain these service when we describe the architecture of the
StratOSphere system.

5. StratOSphere architecture

Figure 6 illustrates the multi-tiered layer of design services that comprise **StratOSphere**.
These layers consist of the Transport layer, the Repository Layer, the Messaging Layer, the

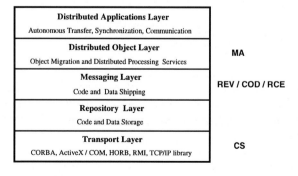

Figure 6. **StratOSphere** design layers.

Object Migration Layer, and the Distributed Application Layer. Annotated next to each relevant layer is the form of mobility provided by the architecture.

5.1. Transport layer

The bottommost layer of **StratOSphere** is the Transport Layer which provides basic remote invocation services for distributed applications. In this layer we take a simple CS system as a substrate upon which we can provide further mobility paradigms. Here we may select from the myriad of packages: RPC, CORBA, DCOM, DCE, and Java-specific interoperability frameworks such as HORB [33] or RMI [44]. Each of these packages permit a client to pass arguments and invoke a method of a remote object, wait for the method to be executed at the server and obtain a result, thus satisfying basic CS functionality.

For the Transport Layer, we have chosen HORB as our CS package to provide remote access and remote execution. Though HORB is capable of much greater functionality, we utilize HORB simply to provide delivery and remote access to objects. Access to these HORB services are restricted to the Transport Layer, so that transport mechanisms can be substituted in place without disturbing the remaining layers of the **StratOSphere** architecture; we could just as well have replaced HORB with any of the other packages above by re-implementing our Transport Layer interface. Indeed, General Magic's Java-based agent and distributed object system, Odyssey [42], lets the application programmer select among RMI, DCOM, and CORBA IIOP for its transport.

5.2. Repository layer

At each site in a distributed system, we require the services of a repository to store data and code, where data and code correspond to *instances* and *external methods* of a **StratOSphere** object. The Java object serializer [30] is employed to serialize each object into a byte stream for storage in a repository. Since an external method is a Java `.class` file, it can be stored in a repository as a stream of Java byte-code. In this way, the repository layer serves as a persistent storage system for mobile objects in **StratOSphere**. By storing serialized instances, we can record the state of an object, as it proceeds in computation. A repository further serves as a library of services by storing external methods that act upon object instances.

To operate within the **StratOSphere** framework, data instances and external methods must satisfy the following Java interfaces:

```
public interface SSObject extends Serializable {
}

public interface SSInstance extends SSObject {
    public boolean invariant();
}

public interface SSMethod extends SSObject {
```

```
    public boolean precondition(SSInstance instance,
            String methodName, SSArglist arglist);
    public SSResult run(SSInstance instance,
            String methodName,
            SSArglist arglist) throws SSException;
    public boolean postcondition(SSInstance instance,
            String methodName, SSArglist arglist);
}
```

Each **StratOSphere** instance implements the SSInstance interface in order to be serialized and stored in a repository, while each external method implements SSMethod in order to be dispatched by the **StratOSphere**'s run-time environment. We defer discussion of the invariant, precondition, and postcondition methods in these interfaces until Section 5.4; for now, we will examine the run method in the SSMethod interface, which is used to dispatch an external method upon an instance.

The implementation of the method dispatcher is described in [45]; in essence, **StratO-Sphere** employs an MSQL database [19] to store instance, class hierarchy, and class format meta-data tables. We wrote a new classloader [14] that queries these tables to locate the appropriate internal or external method to apply upon an object, based upon the name of the method specified in methodName, and the instance class type. If the query resolves to an external method, then the dispatcher uses our classloader to load the external method from the repository, and invoke the run method, passing in the required arguments: the **StratOSphere** instance, the name of the method, and an appropriate argument list. If the query returns an internal method of the **StratOSphere** instance, then we employ the Java Reflections Library [38] to dynamically invoke the internal method of the object instance, again passing in the relevant arguments: the object instance and argument list.

5.2.1. Repository structure. Figure 7 illustrates our repository structure within **StratO-Sphere**. Each site implements both a *Local repository* and a portion of a *Global Repository*.

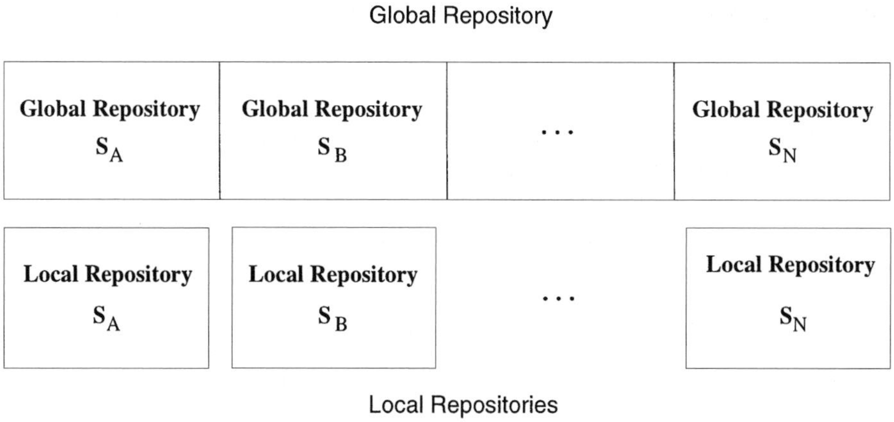

Global Repository

| Global Repository S_A | Global Repository S_B | $\cdots$ | Global Repository S_N |

| Local Repository S_A | Local Repository S_B | $\cdots$ | Local Repository S_N |

Local Repositories

Figure 7. **StratOSphere** repository structure.

49

The Global Repository can be viewed as a global address space partitioned among the hosts in the network; objects stored in the Global Repository are visible at every **StratOSphere** site. Each Local Repository, however, is separate from the others. A Local Repository caches object instances, and provides local implementations of external methods. A discussion on how we maintain consistency among different copies of objects cached at each Local Repository will not be presented in this paper, but we briefly describe the storage of external methods at each Local Repository.

As noted above, our classloader queries a set of meta-data tables at each repository site to locate a particular method to dispatch. Each repository site may, in fact, provide a different implementation of an external method. For example, sites S_A and S_B may each store a different implementation of the segment method in their Local Repositories. A Image object visiting S_A would obtain a different behavior than if it were to visit site S_B. Our **StratOSphere** framework thus supports the *factory pattern* [12], in which a common abstract interface is adopted to obtain different concrete method implementations at each site. A repository site can take advantage of this feature to provide specialized behavior for a particular method. A different segmentation algorithm may be implemented for the segment method, for example, if the host were a supercomputer as opposed to a desktop machine.

In the **StratOSphere** system, our repository structure exhibits both vertical and horizontal partitioning. The Global Repository is vertically partitioned to extend a common storage space across different sites. The Local Repositories are horizontally partitioned to implement new object behavior at each individual site.

5.2.2. Repository scope. The schema for our object repository is shown in figure 8, where we note that instance and method objects each have three levels of scoping: *Global*, *Local*, and *Private*. Global scope is extended to elements in the Global Repository, as it serves as a global address space for the entire network. A site with a Local Repository has Local scope, since its implementations are visible only to itself, and not to any other site. Finally, we come to the concept of Private scope, which is applicable to MA applications. Each **StratOSphere** mobile agent contains a Private Repository that travels with the agent as it visits different sites. We will discuss this notion of Private Repository further in Section 5.4, when we describe our MA implementation. There we shall see that a mobile **StratOSphere** application

	Instance Store	Method Store
Private Scope	**PrivateInstanceStore**	**PrivateMethodStore**
Local Scope	**LocalInstanceStore**	**LocalMethodStore**
Global Scope	**GlobalInstanceStore**	**GlobalMethodStore**

Figure 8. **StratOSphere** repository schema.

can compose different views of a distributed object by visiting different repositories for their local implementations, yet retain access to Private and Global repository objects.

5.3. Messaging layer

The next layer of our design is the Messaging Layer, in which we model interactions as message-passing activity between remote entities. A sender sends a message to a receiver by specifying a remote repository as a location. The message is then delivered to its destination through the Transport Layer using the Repository Layer to reference the target repository.

In this layer, each message passed between remote objects is itself a Java object, and not simply a coded tag or field value. For this layer, we provide an interface called SSDispatchable.

```
public interface SSDispatchable {
    public SSResult dispatch() throws SSException;
}
```

Each message sent from client to server corresponds to a Java object that implements this interface, by defining a dispatch routine. Upon receiving the message, the server simply calls dispatch to process the message. Since the message is an object, the code to execute the message is encapsulated entirely in the message, and need not be registered at the server.

By designing the message-passing mechanism in this manner, we can implement REV and COD functionality for our mobile code system. As noted earlier, REV can be viewed as a natural extension of CS, if executable code were able to be represented as data; in the case of Java this poses no barrier, as external methods in the form of Java programs can be stored as an array of Java byte-code. The byte code can then be inserted into a message along with the serialized object instance, and passed from client to server. When the server processes the message, the executable byte-code is subsequently extracted from the message and executed upon the instance using a **StratOSphere** Java classloader [14]. The result of the execution is sent back to the client by means of a subsequent message.

By factoring out the message-passing interaction between client and server in CS and REV, we are able to build REV functionality on top of the Transport layer's CS paradigm. Likewise, COD can also be implemented in a very similar manner: A message is sent to an server requesting an repository method; the server obtains the byte-code for the external method and passes it back to the client a return message; the client extracts byte-code, dynamically loads and executes the method at the client's site. The manner in which we obtain both REV and COD functionality is through specialization of the messages that are passed between client and server.

In addition to implementing a Dispatchable interface, a message object must provide one other general routine, which is the send method:

```
public abstract class Message implements SSDispatchable {
    public SSResult send() throws SSException {
        ...
    }.
}
```

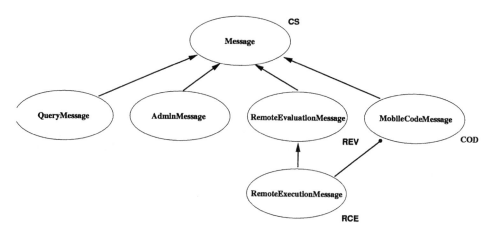

Figure 9. **StratOSphere** messages.

The send method transfers the message object to a given repository site. From this message-delivering abstract Java class, we may then derive other subclasses to provide specialized services.

In figure 9, we see the pertinent message objects in the **StratOSphere** system, with the Message at the root of this hierarchy. A client sends a QueryMessage to locate and explore information at remote sites. An AdminMessage is issued by a privileged client to perform some administrative function such as shutting down the repository site, introducing a new object type into the system, or modifying the contents of an object repository. The RemoteEvaluationMessages, MobileCodeMessages, and RemoteExecutionMessages provide the system with mobile code functionality. Each of these messages holds an SS-RepositoryObject to fetch or store data instances and external methods into the repository. The RemoteEvaluationMessage first fetches data and code from the local repository and delivers the message to the destination site for remote execution, providing the system with REV capability. To satisfy COD, the MobileCodeMessage is sent to fetch objects from the remote repository. Finally, the RemoteExecutionMessage provides RCE functionality by sending the message to the execution site, then issuing MobileCodeMessage to pull the relevant code and data and directly to that site for processing.

The Messaging Layer accesses the services of the Repository Layer through a set of well-known interfaces. To obtain an object in a **StratOSphere** repository, these messages go through the SSRepositoryObject interface,

```
public interface SSRepositoryObject {
      public void bind() throws SSException;
      public void fetch() throws SSException;
      public void store() throws SSException;
}
```

which defines bind, fetch, and store methods.[3] The bind method associates a particular code or data item with a repository; the bind method must always be called before any further operations can be invoked. The `fetch` method is called to extract the item out of repository storage, while the `store method` inserts it back in.

We note that in each of these three messages that the order in which the `send` of the message and the `fetch` of the `SSRepositoryObject` vary. The `RemoteEvaluation-Message` performs a `fetch` of the objects before the `send` to the server site. Conversely, the client must first `send` the `MobileMessage` to the repository site, before it can `fetch` the repository item and return it to the client. Since RCE is a union of REV and COD, the `RemoteExecutionMessage` will first `fetch` any local data before the `send` to the execution site, whereupon the `RemoteExecutionMessage` issues a `MobileObjectMessage` to `fetch` any remaining objects from remote repositories.

With REV, COD, and RCE in place in the Messaging Layer, the **StratOSphere** system can satisfy both local and remote execution scenarios: L1–L4 and R1–R9. To obtain MA functionality, however, we must develop further mechanisms at higher levels of **StratOSphere**.

5.4.   Distributed object layer

The previous Messaging Layer provided a means of shipping code and object instances across remote sites in order to retarget processing of the object. In this layer, we use the services of the Messaging Layer to implement a distributed object, a mobile object that can be dynamically relocated to a different host in a type-safe manner. To do so, we introduce the addition of a global naming service and a distributed execution facility.

5.4.1. Global naming service.   The Messaging Layer provides passing of objects and methods so that objects are copied to remote sites, but to provide distributed processing, we must also have some means of accessing object by reference. For this, we require the addition of a global naming service that keeps track of the location of each object as it is migrated, and maintains a global pointer to the running object.

In our current design for **StratOSphere**, we employ the services of a centralized database to track object state and location. This database has global scope, so it is accessible to all entitles in the system. Each time a **StratOSphere** instance is created, an entry is added into a table, providing the following information:

(ObjectName, ObjectRef, Host, Owner, Type, State, Time).

The *ObjectName* is a string assigned to an object, and used as a primary key for this global name table. The *ObjectRef* is an unsigned integer value that stores a host-specific reference to the object. The *Owner* and *Host* fields describe the hosts where the object was originally created, and where it is currently located. The *Type* field stores the qualified name (full package name) of the Java object, while *State* value describes the current object state (*Created, Transferred, Stored, Running, Destroyed*), corresponding to the states when the object was created, when it was migrated to a new site, when it was checked in and checked out of a persistent store, when an operation was invoked on the object, and when it was destroyed or garbage collected. Finally, the last field *Time* records the global timestamp that the entry was entered into the database.

Using this information, we can keep track of an object's current status and provide access to an object through a global reference. Though our approach is currently based upon a centralized server, we have studied plans to redesign the naming server into a distributed server. The work of Awerbach and Peleg [3], for example, describes the design of a distributed directory for mobile users, that tracks the location of each mobile object in an efficient manner by incrementally propagating updates into the directory for specific subnetwork regions.

5.4.2. MA facility.   The next service that we provide is a means of distributing the execution of an object throughout the network: MA. In the Messaging Layer, we designed a facility to migrate and deliver objects by storing them within appropriate messages, and passing the messages between hosts. The drawback with this technique is that the object migration is specified on a peer-to-peer basis; there is no coordination among different hosts to perform distributed processing of an object. The following example illustrates the need for this facility to distribute execution:

Suppose we have a client application that is logged on a laptop machine named skinny. The client would like to create a LandsatImage of Santa Barbara County, segment the image to locate areas of vegetation growth, and render a high quality hardcopy graphic of the resulting segmented area. The resources to perform this task, however, are scattered and distributed all over the network. A large LandsatImage of California state is stored at a mainframe called bigblue; the most powerful machine on the network is an UltraSparc named speedy, and a dedicated high-resolution graphics device is attached to a dedicated graphics processor called picasso. As we shall see, we rely on the ability to coordinate execution and migration of objects to complete this distributed task in an resource-efficient manner. The distributed processing of this object is illustrated in figure 10. To actually

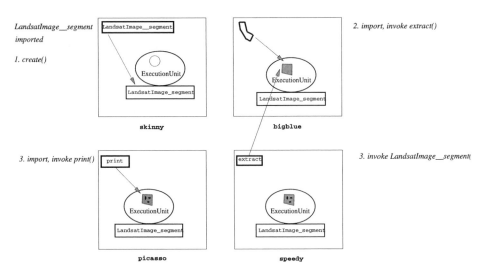

Figure 10.   Remote execution facility over four machines.

implement such an execution, we first define the notion of a migrating object store which we call a *Execution Unit*. The *Execution Unit* holds a Private Repository to store its own collection of instance and external methods, and a instruction queue. By executing each instruction at a repository site, the *Execution Unit* is able to visit remote repositories to gather new data and methods into its Private Repository; the *Execution Unit* can thus compose its own view of an object, aggregated from distributed sources. There are a set of primitive instructions to program a *Execution Unit*; these are: *new, import, export, push,* and *invoke*.

The new command calls a constructor to create an object. The import instruction imports a repository object from a Global or Local Repository and stores the item into its Private Repository, while the export has the opposite effect. The push migrates the *Execution Unit* to a new site, akin to the go command in Telescript and Sumatra, where it resumes execution with the following instruction.

These instructions operate on the object in the following manner:

```
instruction address    operation                              name

  import    {*}         Factory::LandsatImage                  as Landsat
  new                   Landsat                                as image
  import    {skinny}    LandsatImage::LandsatImage__segment    as segment
  push      {bigblue}
  import    {speedy}    LandsatImage::extract                  as extract
  invoke                image.extract()                        as sbimage
  push      {speedy}
  invoke                sbimage.segment()
  push      {picasso}
  import    {picasso}   LandsatImage::print                    as print
  invoke                sbimage.print()
  export    {picasso}   sbimage
```

The client initially creates an empty LandsatImage object by importing a factory class from the Global Repository. The address of * indicates that the directory of the Global Repository is queried to locate the pertinent class. The result is stored in the Private Repository of the *Execution Unit* and assigned a name of *Landsat* within *Execution Unit's* namespace. Next, the *new* instruction creates a LandsatImage object named *image*, again stored within the *Execution Unit*, while still located at skinny.

The client's goal is to segment and print a region of Santa Barbara County, unfortunately, there is no *segment* operation in the LandsatImage class; instead, the client must provide one. The client creates a class called LandsatImage__segment, compiles it into byte-code, and temporarily inserts it into skinny's Local Repository.

```
public class LandsatImage__segment implements SSMethod {
    public class LandsatImage__segment() {
        // SSMethod constructor
    }
```

```
public boolean precondition(SSInstance instance, String
    methodName, SSArglist arglist) {
    // First assert that the instance is of Class
    // LandsatImage to ensure type-safety. Then
    // check to make sure that instance is an non-empty
    // image of appropriate size.
}
public SSResult run(SSInstance instance, String
    methodName, SSArglist arglist) throws SSException {
    // Extract the arguments from the arglist
    // to specify a region to segment and
    // properties to segment.  Then segment
    // the image stored in instance.
}
public boolean postcondition(SSInstance instance, String
    methodName, SSArglist arglist) {
    // Check to make sure that the segmented
    // edges are stored in the SSResult, and
    // the image has not been corrupted.
}
}
```

The method is located from the Local Repository at skinny by specifying its name (*segment*), and the name of the type it must operate upon (*LandsatImage*). Within the Local Repository directory, these two names resolve to the external method *LandsatImage__segment*. Using the *import* command, the client loads the method into the *Execution Unit's* Private Repository for later execution at a more powerful machine; the external method is named *segment* in the *Execution Unit*.

The *push* command then migrates the unit to the mainframe bigblue, where a large LandsatImage of California resides. The extract executable, however, is stored at the UltraSparc speedy, leading to two possibilities: either migrate the California object from bigblue to speedy for remote execution (R3), or import the executable from speedy and perform the operation locally (L2). To migrate the California object, the *Execution Unit* would first have to insert the California LandsatImage into its repository then *push* onward to speedy. Unfortunately California is much too large an object to shift, consuming far too much bandwidth, the client opts for the latter alternative. The *pull* method transfers the extract method from speedy's Local Repository into the *Execution Unit's* repository naming it extract also, and executes the method upon the California object at bigblue. The result of the extraction is a Santa Barbara County LandsatImage which is then stored in the *Execution Unit* as *sbimage*.

Next, the *Execution Unit* proceeds to picasso, where it dispatches the LandsatImage__segment method to mark the areas of vegetation growth in *sbimage*. The *Execution Unit* then *pushes* itself to picasso, where it *imports* a specialized print external method from picasso's Local Repository, and then *invokes* this *print* command to produce a hardcopy. Finally, the unit *exports* the current sbimage, transferring it from its Private Repository to the current Local Repository.

At this point, we briefly describe how we ensure correctness and type-safety as it appears we are permitted to dispatch any external method upon any instance, without regard to object semantics. Earlier, we noted that the SSInstance interface specified an invariant method, while the SSMethod interface required both precondition and postcondition methods. These assertion checks at run-time [24] provide us with a means of checking type-safety and ensuring the correctness of each operation. Each **StratOSphere** object specifies a class invariant, which our classloader checks whenever it dispatches a method, to ensure that the object state does not become inconsistent. For example, our LandsatImage has the following invariant:

```
public class LandsatImage implements SSInstance {

    private Dimension dimension;
    private Image       image;

    public boolean invariant() {

        // Check to make sure dimensions are valid,
        // and image has not been corrupted.
    }

    public class LandsatImage() {

        // SSinstance constructor
    }
}
```

After each external method has executed, the classloader checks the invariant to ensure that the LandsatImage has not been corrupted.

Likewise, our classloader checks each SSMethod's preconditions and postconditions before and after each external method dispatch. In our LandsatImage__segment method above, the precondition ensures that the SSInstance is of the appropriate type by calling the Java getClass method upon the SSInstance to determine its actual type; the precondition further checks for correctness by requiring that the method does not encounter an unexpected instance state. The postcondition then ensures that the routine performs as specified, and that the segmented edges are then stored in SSResult. If any of the precondition, postcondition, or invariant assertions are violated, the classloader triggers an SSException which is returned to the client application with to report the interface assertion violation.

The *Execution Unit* provides the functionality that we require to coordinate distributed processing by incorporating both an internal private object repository and an instruction queue. The *Execution Unit* interprets the instructions at run-time; it performs these *import* and *export* instructions, by issuing MobileCodeMessage and RemoteEvaluationMessages to transfer instances and methods between repositories using COD and REV. The **StratOSphere** classloader is employed to dispatch these methods and instances to invoke a method. Similarly, the *new* instruction is processed by calling the appropriate Java object's class constructor.

The remaining *push* instruction also uses REV to transfer the *Execution Unit* to a remote site. In doing so, however, a few intricacies arise: We assume that the state of each **StratOSphere** object is safely stored in the Private Repository, so that the object can continue processing after a migration; we must also ensure that the state of the *Execution Unit* itself is maintained so that after a *push*, the subsequent instruction in the instruction queue is executed. To resolve this problem, we subclass the *Execution Unit* from the Remote-ExecutionMessage in order to deliver it to remote repository sites for further processing, satisfying execution scenarios, L1–L4 as well as R1–R9.

6. Alexandria data model

As geographical information constitutes the bulk of data in the Alexandria Digital Library, most of the objects that we encounter in the Data Store model geographical data types [31]. We now describe the Data Store repository, which provides a dynamic and extensible type system for objects in the Data Store collection.

To motivate the discussion, we use as an example the class hierarchy of figure 11. We briefly describe each data type presented in the figure: LandsatImages are high altitude, medium resolution images that are useful for terrain identification and analysis; HyperSpectralImages have hundreds of spectral channels and allow objects in the image to be identified by their component spectra; RadarImages are active-sensored data used for all-weather terrain imaging; TigerMaps combine maps of streets, cities, and counties along with census data to produce population and traffic studies; DLGMaps (Digital Line Graphs) present layers of transportation, hydrographic, and urban planning data for earth modeling and simulation.

To model the organization and relationships among these different types, we construct the following *Instance*, *ClassFormat*, and *ClassHierarchy* relations in Tables 4–6. Each entry in the *Instance Table*, consists of an instance name and a class name, corresponds to an object in the Data Store. The instance name refers to some static representation of the Data Store object. In the case of images, this representation is the bitwise pixel map of the image. A Data store object is created by first instantiating a Java-program object and then

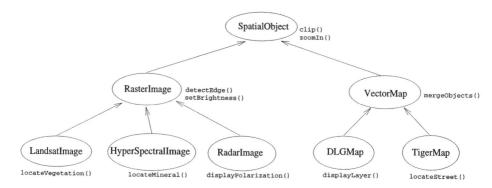

Figure 11. DataStore class hierarchy.

Table 4.   Instance table.

instanceName	className
SB-COUNTY-97-04-01	LandsatImage
GOLETA-CITY-95-03-02	LandsatImage
UCSB-CAMPUS-96-05-07	DLGMap
SOLVANG-CITY-97-12-12	TigerMap

Table 5.   ClassHierarchy table.

class	parent
SpatialObject	Root
VectorMap	SpatialObject
RasterImage	SpatialObject
LandsatImage	RasterImage
HyperSpectralImage	RasterImage
RadarImage	RasterImage
DLGImage	VectorMap
TigerImage	VectorMap

Table 6.   ClassFormat table.

class	methodName	storageName
SpatialObject	zoomIn	SpatialObject__zoomIn
RasterImage	zoomIn	SpatialObject__zoomIn
RasterImage	detectEdge	RasterImage__detectEdge
VectorImage	zoomIn	VectorImage__zoomIn
VectorMap	mergeObjects	VectorMap__mergeObjects
LandsatImage	zoomIn	SpatialObject__zoomIn
LandsatImage	detectEdge	RasterImage__detectEdge
LandsatImage	locateVegetation	LandsatImage__locateVegetation
TigerMap	zoomIn	VectorImage__zoomIn
TigerMap	mergeObjects	VectorMap__mergeObjects
TigerMap	locateStreet	TigerMap__locateStreet

calling a Data Store constructor that converts the Java-level object into a Data Store object, initializing it by loading in its static representation.

The *ClassFormat Table* describes the behavior of each class by listing the available methods in the class; every method that a Client specifies in a RemoteExecutionMessage must have a corresponding entry in the methodName column. Associated with the methodName is the location of the program code that the Server invokes on behalf on the Client to execute

59

the operation. Run-time type checking information in the program code ensures that the appropriate method be applied upon an object for consistent execution.

The remaining table, the *ClassHierarchy Table*, describes the inheritance relationships among the Data Store classes. Just as in the usual object-oriented languages, implementation inheritance is supported in our model. Since a `TigerMap` is-a `VectorMap`, it inherits the `mergeObjects` method from `VectorMap` by re-using the `VectorMap__mergeObjects` program code for its method implementation.

These *Instance*, *ClassFormat*, and *ClassHierarchy* tables provide object management facilities for the Data Store. A Client application can issue database queries (via a `QueryMessage`) to browse the Data Store repository for meta-data and identify objects of interest. Upon locating a particular object representation, the Client instantiates the object dynamically by sending a `MobileCodeMessage` to retrieve the object to the Client site.

In designing this data model, we have provided a schema that adheres to a relational model, in order to facilitate storage in a relational DBMS. Other approaches to decomposing and mapping objects into relational databases have been explored in Postgres [37].

By providing the basic structure and methods of the Data Store in terms of a DBMS that allows Clients to dynamically execute methods, we no longer need a statically defined proxy for each object in the Data Store. Rather, the Client can dynamically discover relevant methods to execute on the objects of interest.

We note here the departure from the usual approach adopted by distributed object systems such as CORBA. Each CORBA distributed object is compiled into a proxy object, which in turn acts as a stub to provide remote access for the desired object. In the Data Store, there is no need to generate proxies for a Data Store object. A new object type such as `WeatherMap` can be introduced into the Data Store without having to provide a new proxy. We need only add new `WeatherMap` entries into the *Instance*, *ClassFormat*, and *ClassHierarchy* tables, and supply the required method programs that implement `WeatherMap` operations.

The benefit of this dynamic object access lies in the run-time time interpretation of the Data Store object model. The class format and inheritance relations of `LandsatImage` are defined statically. Once instantiated, a Java `LandsatImage` remains fixed. There are no provisions to extend or alter the `LandsatImage` class model within the Java language. Instead, we provide this dynamic extension through the Data Store's own data model.

7.   Conclusion and future work

Work on agent security has been an area of active research [17]. Among the issues involved are scenarios in which a malicious agent may attack a host, a host may attack an agent, or an agent may attack another agent. In our design, we have greatly reduced this problem by assuming that each host repository is a trusted source; that is, only privileged administrators may maintain and update the repository. Consequently, we assume that a host will not attack an agent, but concern ourselves with the more scurrilous problem of an agent attacking the host. In our initial implementation of the Distributed Applications Layer, we use signed certificates [39] to authenticate each agent. No host or agent exchanges any data or information with another agent unless a signed certificate identifies the agent as a trusted source. Another approach that we are investigating includes proof-carrying code [25], which performs a verification of code to be executed to ensure trusted operation.

The final layer of interoperability in **StratOSphere** is the Distributed Applications Layer, from which distributed applications can be built with MA execution. This is the level where mobile code languages systems such as Aglets [23], TACOMA [21], Telescript [43], Odyssey [42], and Voyager [26] provide services for enterprise applications. At time of writing, this final layer is still being developed and implemented for **StratOSphere**. By utilizing the services of the Distributed Object Layer, we have the migration capability of an MA system. Designing this layer, however, involves not just a mobility paradigm, but also issues of inter-agent communication, synchronization, and agent security.

Inter-agent communication varies among different agent systems. In our design, we use an idea initially developed in MØ [40], in which agents exchange data in a *shared memory* area of each execution site. In adapting this idea to **StratOSphere**, we have reserved a portion of each Local Repository, called the *Exchange Area* to provide both data exchange and code exchange among different agents. Synchronization of updates to this Exchange Area is supported by a thread queue for each mobile process.

After studying all possible execution scenarios, we have designed and built a framework called **StratOSphere** that distributes processing across multiple host sites, and also defines a dynamic object model to implement client services. By building a multi-tiered architecture, we were able to successfully develop a dynamic and extensible distributed environment for **StratOSphere** objects, providing REV, COD, and RCE in the Messaging layer and MA in the Distributed Object Layer. We show how the object instances and object methods can be selectively migrated to different sites for remote processing. We describe the **StratOSphere** naming service and object repository, and provide a programming model to build a dynamic *Execution Unit* in which sequences of method executions are targeted at remote sites for distributed processing. We have developed a prototype of the **StratOSphere**, and are currently investigating other applications involving MA and distributed database queries in the distributed application layer using the *Execution Unit*.

Notes

1. Indeed, Xerox PARC's variant of CORBA, ILU [20], mixed-language programming is supported to such an extent that ILU objects implemented in different languages can reside in the same process space.
2. See CORBA's Common Object Services Specification [35], DCE's object services [32], and DCOM library services [5] for further details.
3. For sake of simplicity, formal arguments of methods will not be shown in the text.

References

1. A. Acharya, M. Ranganathan, and J. Saltz, Sumatra: A Language for Resource-aware Mobile Programs, Springer Verlag Lecture Notes in Computer Science, 1997, pp. 111–130.
2. K. Arnold and J. Gosling, The Java Programming Language, Prentice-Hall: Reading, MA, 1996.
3. B. Awerbuch and D. Peleg, "Online tracking of mobile users," Journal of the Association for Computing Machinery, vol. 42, pp. 1021–1058, 1995.

4. A.D. Birrell and B.J. Nelson, "Implementing remote procedure calls," in Proc. ACM Symp. on Transactions on Computer Systems, February 1984, pp. 19–59.
5. D. Box, Creating Components with DCOM and C++, Addison Wesley, Longman, 1997.
6. L. Cardelli, "A language with distributed scope," in Proc. of the 22nd ACM Symposium on Principles of Programming Languages, 1995.
7. A. Carzaniga, G.P. Picco, and G. Vigna, "Designing distributed applications with mobile code paradigms," in Proc. of the 19th Intl. Conf. on Software Engineering, 1997.
8. P.I. Chang, Inside the Java Web Server, Javasoft, Inc., http://java.sun.com/-features/1997/aug/jwsl.html, 1997.
9. D.R. Cheriton, "The v distributed system," Communications of the ACM, pp. 314–333, March 1988.
10. G. Cugola, C. Ghezzi, G.P. Picco, and G. Vigna, "A characterization of mobility and state distribution in mobile code languages," in Proc. of the 2nd Workshop on Mobile Object Systems, July 1997.
11. F. Douglis and J. Ousterhout, "Process migration in the sprite operating system," in Proc. of the 7th Intl. Conf. Distributed Computer Systems, 1987, pp. 18–25.
12. E. Gamma, R. Johnson, R. Helm, and J. Vlissides, Design Patterns: Elements of Reusable Object-Oriented Software, Addison-Wesley, 1995.
13. A. Goscinski, Distributed Operating Systems: The Logical Design, Addison-Wesley, 1991.
14. J. Gosling, B. Joy, and G. Steele, The Java Language Specification, Sunsoft Java Series, Addison Wesley, 1996.
15. R.S. Gray, "Agent tcl: A flexible and secure mobile-agent system," in Proc. of the 4th Annual Tcl/Tk Workshop, 1996, pp. 9–23.
16. S. Gundavaram, CGI programming on the World Wide Web, O'Reilly & Associates: Cambridge, 1996.
17. J.D. Guttman, W.M. Farmer, and V. Swarup, "Security for mobile agents: Authentication and state appraisal," in Fourth European Symposium on Research in Computer Security Proceedings, September 1996, pp. 118–130.
18. J. Hartman, U. Manber, L. Peterson, and T. Proebsting, "Liquid software: A new paradigm for networked systems," Technical Report Technical Report 96-11, University of Arizona, June 1996.
19. D.J. Hughes, "Mini sql: A lightweight database server," http://Hughes.com.au/-product/msql, 1995.
20. B. Janssen and M. Spreitzer, ILU 2.0 Reference Manual, Xerox PARC, http://-ftp.parc.xerox.com/pub/ilu/ilu.html, 1996.
21. D. Johansen, R. van Reneese, and F. Schneider, "An introduction to the tacoma distributed system," Technical Report Technical Report 95-23, University of Tromso, 1995.
22. E. Jul, H. Levy, N. Hutchinson, and A. Black, "Fine-grained mobility in the emerald system," in Proc. ACM Symp. on Transactions on Computer Systems, 1988, pp. 109–133.
23. D.B. Lange and D.T. Chang, "Ibm aglets workbench: Programming mobile agents in java," http://www.trl.ibm.co.jp/aglets/whitepaper.htm, 1996.
24. B. Meyer, Object-Oriented Software Construction, Prentice-Hall, 1988.
25. G.C. Necula and P. Lee, "Safe kernel extensions without run-time checking," in Proc. of the 2nd ACM Symposium on Operating System and Design and Implementation, October 1996.
26. ObjectSpace, "Voyager technical overview," http://www.objectspace.com/voyager/-technical_white_papers.html, 1996.
27. Open Software Foundation, Introduction to OSF DCE: Rev 1.0, Prentice Hall: Engle-wood Cliffs, NJ, 1992.
28. G. Popek and B.J. Walker (Eds.), The Locus Distributed System Architecture, MIT Press, February 1986.
29. J. Richardson and P. Schwarz, "Aspects: Extending object to support multiple independent roles," in Proc. ACM SIGMOD Int. Conf. on Management of Data, May 1991, pp. 298–307.
30. R. Riggs, J. Waldo, and A. Wollrath, "Pickling state in java," in Second Conf. on Object-Oriented Technologies and Systems (COOTS), Toronto, Ontario, June 1996, pp. 241–250.
31. A.H. Robinson, Elements of Cartography, Wiley: New York, 1995.
32. W. Rosenberry, D. Kenney, and G. Fisher, Understanding DCE, O'Reilly & Associates: Sebastopol, CA, 1992.
33. H. Satoshi, The Magic Carpet for Network Computing: HORB Flyer's Guide, Electrotechnical Laboratory, http://ring.etl.go.jp/openlab/horb, 1996.
34. G. Semeczko and S.Y.W. Su, "Supporting object migration in distributed systems," in Proc. of the Fifth Intl. Conf. on Database Systems for Advanced Applications, Melbourne, Australia, April 1997, pp. 59–66.

35. J. Siegal, CORBA: Fundamentals and Programming, Wiley, 1996.

36. T.R. Smith and J. Frew, "Alexandria digital library," Communications of the ACM, vol. 38, no. 4, pp. 61–62, 1995.

37. M. Stonebraker, "Object management in postgres using procedures," in 1986 International Workshop on Object-Oriented Database Systems, Pacific Grove, Calif., September 1986, pp. 66–72.

38. Sun Microsystems, Inc., Java Core Reflection API and Specification, http://java.sun.com/products/jdk/1.1/docs/guide/reflection/-index.html, 1997.

39. J. Tardo and L. Valente, "Mobile agent security and telescript," in 41st TEEE Computer Society Intl. Conf., February 1996, pp. 58–63.

40. C. Tschudin, The Messenger Environment MØ—A Condensed Description, Lecture Notes in Computer Science, Springer Verlag, 1997, pp. 149–156.

41. P. van Roy, S. Haridi, P. Brand, et al., "Mobile objects in distributed oz," in Proc. ACM Symp. on Transactions on Programming Languages and Systems, September 1997, pp. 804–851.

42. J. White, "Mobile agents white paper," http://genmagic.com/agents/Whitepaper/-whitepaper.html, 1996.

43. J. White, "Telescript technology: Mobile agents," http://genmagic.com/TeleScript/-WhitePapers, 1996.

44. A. Wollrath, R. Riggs, and J. Waldo, "A distributed object model for java," in Second Conf. on Object-Oriented Technologies and Systems (COOTS), Toronto, Ontario, June 1996, pp. 219–231.

45. D. Wu, D. Agrawal, A. El Abbadi, and A. Singh, "A java-based framework for processing distributed objects," in Proc. Intl. Conf. on Conceptual Modeling, Los Angeles, CA, 1997, pp. 333–346.

Distributed and Parallel Databases, 7, 319–342 (1999)

Experiences of Using Generative Communications to Support Adaptive Mobile Applications

ADRIAN FRIDAY
NIGEL DAVIES   adrian@comp.lancs.ac.uk
JOCHEN SEITZ, MATT STOREY AND STEPHEN P. WADE
Distributed Multimedia Research Group, Computing Department, Lancaster University, Bailrigg, Lancaster, UK

Abstract. Attention has recently begun to focus on the use of asynchronous paradigms to support adaptive mobile applications. To investigate this issue the authors have developed an asynchronous distributed systems platform based on the tuple space paradigm [19] coupled with extensions to support operation in mobile environments. This paper presents our experiences of developing and using this platform. The benefits of the tuple space approach are highlighted and we discuss in some detail the design, implementation and performance of our platform. We subsequently focus on the critical issues of the tuple space API and the level of support for adaptation which can be provided without compromising the elegance and simplicity of the paradigm. The paper concludes with an analysis of the suitability of platforms based on the tuple space paradigm for use in mobile environments.

Keywords: mobile computing, adaption, tuple spaces

1. Introduction

Mobile computing environments are characterised by change [10]. More specifically, in an environment in which users and end-systems are highly mobile the resources available to application and system components are subject to rapid and significant fluctuations. For example, an end-system which roams between different network overlays [26] may experience changes in network quality-of-service (QoS) which reduce the effective bandwidth by several orders of magnitude. Furthermore, these changes in resource availability may be mirrored by changes in service availability and, most crucially, user requirements as the context within which the user and end-system operates changes. In order to enable systems to continue to operate in such dynamic environments, it is now widely accepted that system and application components must be *adaptive* [9, 25], i.e., they must be able to adapt their behaviour in response to changes in their context.

Initial approaches to supporting adaptation have focused on one of two techniques: either extending existing distributed systems platforms to enable applications to obtain feedback on network QoS or using proxies to perform adaptation on behalf of applications. The first approach typically involves implementing new APIs which can be used to selectively remove aspects of network transparency and thus expose QoS information. Such extensions are often combined with further refinements such as message buffering which allow applications to continue operation during periods of network disconnection. This approach is typified by the work of Joseph et al. on the Rover system [24]. In contrast, approaches based on proxy architectures [16, 32, 35, 36] allow the instantiation of filtering, caching or translation components into the communications path between clients and servers. Component instantiation is typically carried out either as part of the overall system configuration (usually explicitly by the user or a system administrator) or may be triggered transparently in response to pre-determined or pre-configured QoS events.

More recently, attention has begun to focus on the use of asynchronous paradigms to support adaptive mobile applications. At Lancaster we have developed an asynchronous distributed systems platform based on the tuple space paradigm [19] but extended to support operation in mobile environments. The platform has been fully implemented and used to support a wide range of applications, both mobile and fixed. In this paper we report on our experiences of developing and using the platform. In particular, we focus on the critical issues of the tuple space API and the level of support for adaptation which can be provided without compromising the elegance and simplicity of the paradigm.

Section 2 presents an analysis of the theoretical benefits and shortcomings of the tuple space paradigm when applied to the field of mobile computing. Section 3 then describes in detail the computational and engineering models for our platform. We then consider the API of our platform and discuss support for adaptation in tuple space based architectures (Sections 4 and 5, respectively). Finally, we analyse our experiences and comment on the suitability of platforms based on the tuple space paradigm for use in mobile environments.

2. An analysis of the tuple space approach

2.1. Overview

The tuple space paradigm was conceived in the mid-1980s by researchers at Yale University as a mechanism for coordinating the numerous processes involved in complex parallel computations [18]. A tuple space is an abstract entity, akin to distributed shared memory, spread across all participant processes and/or hosts. Interprocess communications are conducted exclusively through the generation of *tuples* and *anti-tuples* which are submitted to tuple space. This is termed *generative communication*.

Tuples are typed data structures, each formed from a collection of typed data fields, and every tuple represents a cohesive piece of data. Tuples are comparable to structures (structs) in the C programming language, or objects in languages like C++ and Java. Each tuple field is termed either *actual* or *formal*. Actual parameters have both a defined type and value while formal parameters have a defined type but no value, c.f. NULL pointers in C. Tuples of this form are classed as *passive tuples* and are composed entirely of actual and/or formal fields from the time of their creation though to that of their destruction. A second class of tuples, referred to as *active tuples*, is also defined by the tuple space model. In an active tuple, one field or more is of neither an actual or formal nature at the time of creation. Instead, such fields are wholly defined by functions which require evaluation. When an active tuple is deposited in tuple space, a separate process is spawned to calculate each such field and, over time, the tuple evolves into a passive tuple.

Both active and passive tuples are persistent objects and cannot be altered by user processes while they reside in a tuple space. In order to effect changes to a tuple it must be explicitly withdrawn, changed and then re-inserted [19]. Placing a tuple in a tuple space is analogous to transmitting a packet in an IP network and the packet remaining there until the recipient interface chooses to consume it.

Anti-tuples are the antithesis of tuples. While tuples embody a piece of data which has been submitted to a tuple space, anti-tuples capture requests seeking to remove or copy data from the tuple space. In common with passive tuples, anti-tuples are composed from an arbitrary mix of actual and formal fields but instead of defining a piece of data, an anti-tuple contains a template against which to match tuple data. There are two flavours of anti-tuple: *destructive* and *non-destructive*. Destructive anti-tuples each seek to remove a matching tuple from the tuple space. In contrast, non-destructive anti-tuples merely are satisfied by making a copy of a matching tuple, thus leaving the original tuple unaffected.

Tuples are matched against anti-tuples, and vice-versa, by comparing their types, fields, and the values contained in such. Actuals match either formals of the same type, or actuals of the same type and value. Formals only match actuals of the same type. When all the fields of a tuple satisfy an anti-tuple a successful match is made and the formals present in the anti-tuple are replaced by the respective actuals in the matched tuple.

The process of matching of tuples to anti-tuples is performed non-deterministically and, as such, where multiple tuples that satisfy an anti-tuple are available, an arbitrary choice is made. Similarly, if multiple destructive anti-tuples are in existence that match a newly deposited tuple, one of these if chosen at random and satisfied. The tuple space model guarantees the unique withdrawal of tuples, so each tuple can satisfy at most one destructive anti-tuple. Each tuple may, though, rightfully service an infinite number of non-destructive anti-tuples prior to its departure from tuple space.

In itself, the tuple space paradigm is an abstract concept, so the researchers at Yale embodied the paradigm in a coordination language named Linda [18]. Linda was designed with the goal of offering the same combination of simplicity and power to distributed domains that C had offered systems programmers in the sequential one. As such the language was defined with just four simple primitives:

- `eval`: deposits a *active* tuple in a tuple space
- `in`: seeks to *withdraw* information matching a query from a tuple space by submitting a *destructive* anti-tuple
- `out`: deposits a *passive* tuple in a tuple space
- `rd`: seeks to *copy* information matching a query from a tuple space by submitting a non-*destructive* anti-tuple

2.2. *Specific benefits*

Because tuples remain in tuple space from the time of insertion to that of their destructive consumption, those processes producing and consuming tuples need not co-exist. This property is known as *temporal decoupling*. Indeed, once they have been deposited in tuple space, tuples can be consumed by a client at any time, even after the demise of the server(s) which generated them. Therefore, clients and servers can interact using a tuple space without needing to be synchronised with each other. This also means that instead of merely generating tuples in response to queries, producers can create tuples when the data they are to contain becomes available. For example, a weather station which produces hourly reports may deposit a new report tuple into tuple space once each hour, thus obviating the need to repeatedly issue the same information to clients.

Tuple space communications are, by default, *anonymous*, meaning that the client and server are normally unaware of each others identity.[1] As a result of this, coupled with the above temporal decoupling property, there is no implicit requirement to form bindings between client and server processes. Indeed, in those instances where the producer and consumer of tuples do not co-exist, it would not be possible to do so. Thus, the tuple space model has a second important property, that of *spatial decoupling*. Producers simply deposit tuples into tuple space knowing that zero or more processes may access them over time. Because there is no binding, any suitable process can service a tuple which satisfies a request. Therefore, any reachable server that is capable of dealing with the request can destructively withdraw the associated tuple, perform the service and generate result tuples as appropriate.

The spatial decoupling property also enables tuple spaces to provide transparent support for group interactions. A tuple produced by one process can be read by multiple clients in parallel if they all access it using non-destructive anti-tuples. Consider, for example, a groupware application like the shared whiteboard *wb* [15]. The whiteboard process could output tuples describing each drawing operation which are read and rendered by the other whiteboard processes. Through the persistent nature of the tuple space, clients (whiteboards) could join and leave the tuple space at will, with automatic state reconciliation. This greatly contrasts with existing groupware that is often based on RPC based paradigms which require additional, sophisticated group management protocols [6].

2.3. Significant shortcomings

2.3.1. Application programmers interface.
While the temporal decoupling of the tuple space model supports asynchronous communications between processes, Rowstron noticed that the operators which clients use to retrieve data from the tuple space are actually of a synchronous nature [31]. Specifically, while `out` asynchronously deposits data in the tuple space, the `in` and `rd` operators used to retrieve such are blocking and, thus, implicitly synchronous. This means that wherever a tuple matching a despatched anti-tuple is unavailable, the client becomes blocked. In a large number of distributed applications, this behaviour is undesirable. Furthermore, the problem would be further exacerbated in mobile environments where network partitions and disconnection would mean matching tuples remaining unknown, and hence unavailable, for particularly lengthy times.

An additional, well known problem with the tuple space paradigm is often referred to as the 'multiple-rd problem' [30]. In essence, since `rd` requests are matched non-deterministically it is not possible for an application to be sure that it has read in all the tuples which match a given anti-tuple. If these semantics are required, the application must `in` each tuple in turn (removing it from the tuple space) and then `out` the tuples once no more matches are found. While this provides a mechanism for achieving the desired results it places a significant burden on both the application programmer, who must write the appropriate code fragments, and the system which must support many, usually costly, `in` operations.

2.3.2. Supporting adaptation. Mobile computing applications are subject to rapid and significant changes in the QoS available from their supporting environment. Previous research has shown that mobile computing platforms should provide a QoS management architecture that facilitates adaptation, either through application level adaptation or the installation of proxies [17, 32]. However, these approaches have typically focused on communications related QoS, which has, logically, been measured and controlled using bindings. Since the tuple space model does not expressly employ bindings, these existing QoS management architectures cannot be readily integrated into the model in order to support its use in mobile arenas.

3. L²imbo

3.1. Objectives

Given the potential benefits of the tuple space paradigm a number of researchers have attempted to build tuple space platforms based on Linda which target distributed processing over general purpose local-area workstation clusters [5, 22, 27, 29, 33]. These platforms have generally been designed with a very specific application focus and are not considered general purpose tools for building distributed applications. In our work however, we have concentrated on building a general purpose distributed systems platform (called L^2imbo), built on the tuple space concept. Unlike existing distributed tuple space platforms, L^2imbo has been designed to address typical distributed systems issues and function in heterogeneous hardware and software environments. Along with classic distributed systems issues such as portability and support for heterogeneous environments, L^2imbo has been designed to offer a lightweight, minimal set of services (communication, network and architecture transparency) whilst also addressing issues highlighted by existing distributed tuple space implementations, namely, performance and scalability.

One of the key objectives in the design of the L^2imbo platform was to utilise the implicit time and space decoupling of the tuple space platform to facilitate network transparency. We believe traditional platforms suffer in mobile environments because application components must interact via directed synchronous communication. In a typical mobile (and thus currently failure prone) environment a mobile distributed systems platform has two new issues to address: firstly, in the event of failure the communication must be recovered (witnessed by approaches such as Rover's Q-RPC [23]). Secondly, in heterogeneous networks consisting of wired and wireless components, platforms are required to adopt mechanisms to cope with the characteristics of each type of network on the communication path (e.g., M-RPC [3] or Snoop-TCP [1]). However, we believe that with a tuple space based platform in which communications are undirected (mediated by the tuple space itself) it should be possible to deal with the network characteristics on a hop-by-hop basis. Moreover, since the integrity of the tuple space is the sole concern of the platform, one or more networks may be utilised (even simultaneously) without any implications to the platform services offered to the applications. This feature of the L^2imbo platform is currently unique among distributed systems platforms.

In the following sub-sections we examine the computational and engineering models offered by the L^2imbo platform in more detail. In particular, we focus on the protocol architecture and techniques which allow the platform to achieve performance comparable to conventional distributed systems platforms.

3.2. Computational model

Our distributed systems platform provides the same basic API and features as the original Linda model [18] but has been developed to include a number of key extensions:

(i) *Extensions to the API to support asynchronous operations*: We have extended the L^2imbo API using operations based on the Bonita primitives proposed in [31]. These enable clients on each host to access tuple spaces asynchronously by replacing the in and rd operations by two separate operations: one to initiate the operation and one to collect the results at some later time. A further operator allows clients to poll their tuple space interface asking whether the results for a previous request are available. Asynchronous access to the tuple space can both simplify application code and structure (requiring less multi-threading to avoid blocking operations) and may yield improved performance.

(ii) *Multiple local, distributed and centralised tuple spaces which may be specialised for application level requirements such as consistency or security*: The original Linda model allowed only a single shared tuple space (abstracting over a shared area of memory). To address issues of performance, partitioning and scalability, L^2imbo allows the creation of multiple tuple spaces. A tuple space may be one of three basic classes; local (private to that host), distributed (cached at one or more hosts) and centralised (maintained on a single host but accessible from elsewhere). In addition, tuple spaces may be linked using bridging agents which copy tuples between tuple spaces based on factors such as tuple types and QoS parameters [4].

(iii) *System agents which provide services such as tuple space creation, tuple type management, propagating tuples between tuple spaces and QoS monitoring*: All system operations are provided by system agents which the clients interact with using the standard tuple space API. A special set of *system tuples* are used to interaction with the platform agents.

More details of our computation model can be found in [4, 11, 13].

3.3. Engineering model

The L^2imbo platform has been split into two parts: a small stub library, which is compiled into each executable and a daemon process. The stub library maps the actual API calls to the application's host language (currently bindings for C, C++ and Java exist). The current version of the daemon process (version 1.3) is written solely in C and has been ported to run on SunOS 4.1.4 (MULTICAST 4.1.4), Solaris 2.5, Linux 2.0.30 and Windows NT 4.0.

3.3.1. The L^2imbo daemon process.

The L^2imbo daemon process is built using a modular architecture loosely grouped into four layers (see figure 1). The uppermost layer interfaces to the API stub library and is responsible for all communications between the L^2imbo daemon and client processes running co-located on the same host. As access to all tuple spaces is via the daemon process, tuples and anti-tuples generated by separate applications on the same host can be matched without generating network traffic. In addition, the platform is able to gauge the demands on the available network (or networks) of tuple space based applications. This information enables the platform to manage congestion and load balancing more effectively, but incurs a performance penalty since each message involves the overhead of additional local inter-process communications and a context switch.

Below the API interface layer are the tuple space management protocols. In the current implementation only two protocols have been implemented, the distributed tuple space module and the local tuple space module. The distributed tuple space module is responsible for managing a 'conceptually centralised' tuple space between a collection of collaborating platform nodes. We shall consider the distributed tuple space protocol in more detail in Section 3.3.2. The local tuple space module is optimised for access by local applications only and is typically used both for local application inter-process communication and for passing QoS information between the layers of the platform and local applications (see Section 5).

The two lower layers of the platform are regarded as a network abstraction layer which collectively offer a set of transport services. The transport services remain largely independent of both tuple spaces and network technology. The network scheduler accepts protocol messages from the tuple space protocols and, based on associated priority and deadline QoS parameters, determines the order in which they are transmitted. Within each priority, messages are scheduled in earliest deadline first (EDF) order. Messages with the highest

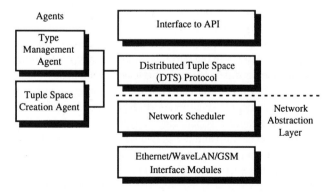

Figure 1. Structure of the L^2imbo daemon process.

priority (lowest number) are considered most urgent and scheduled before those of successive priorities (even if a lower priority has an earlier deadline). This concept is based on previous work by Nieh on thread scheduling for continuous media applications [28].

The lowest layer of the platform consists of a set of packet driver modules for each supported type of network. Each driver module presents a generic interface to the platform behind which details such as connection management and signalling are hidden (although detailed feedback is available via the local tuple space). Packets ready for transmission are delivered to an appropriate network interface module by the network scheduler.

3.3.2. The distributed tuple space protocol. The distributed tuple space protocol module of the platform has been designed specifically to address the issues of scalability and availability of our tuple spaces. In particular, we wanted to avoid solutions which introduced points of failure (poor for networks with mobile elements) or consistency mechanisms based on acknowledgements or token passing (which would degrade performance unacceptably through artefacts such as acknowledgement implosion or the protracted latency of contacting all of the group members).

In implementing the protocol we have chosen to taken advantage of the recent deployment of IP multicast together with application level framing concepts borrowed from work on Scalable Reliable Multicast (SRM) which underpins *wb* [15] and *Jetfile* [20].

The DTS protocol consists of nine distinct protocol messages which are used in conjunction with a cache of tuples (OUT) and anti-tuples (IN) held on each host. Collectively, these caches represent the state of the tuple space. The messages are used to ensure timely propagation of tuples and anti-tuples between caches. An overview of the operation of the protocol is given in Table 1 and the interested reader can find more details in [11].

In order to retain the semantics of the tuple space, it is essential that tuples are never duplicated: once a tuple is injected into the tuple space it must remain unique and must only be withdrawn once by a single process. In existing distributed tuple space approaches this property is maintained by assigning or hashing certain tuples to particular network nodes [14]. This approach would be undesirable in our network environment. Therefore, to ensure the uniqueness property in our platform without penalising system performance, we have introduced the concept of 'tuple ownership'. A tuple can only be removed from the tuple space by its designated owner. The initial owner of a tuple is normally the host which creates it, although the ownership can be reassigned using a message exchange (see CHOWN_REQ and CHOWN_ACK in Table 1) with the current owner of the tuple. By observing sequences of interactions, the platform can determine if a tuple is probalistically likely to be consumed by the originator of the last tuple and can choose to nominate that host as the owner of new tuples it generates. Owner nomination allows RPC-like semantics to be modelled with greater efficiency. Note that since `rd` operations copy tuples non-destructively, they need not be concerned with tuple ownership and hence can be satisfied more quickly and efficiently as they require less communication. In particular, groupware applications in which the same tuple is obtained using `rd` by a number of hosts are supported highly efficiently.

As with the SRM protocol, our platform relies on a 'local repair' mechanism for ensuring that eventual consistency of the tuple space is achieved and maintained. Each protocol module maintains a cache of tuples which have been *snooped* from the multicast group applying

Table 1. Distributed tuple space (DTS) protocol messages.

Message	Format and actions
OUT	`[tuple_id, owner_id, type, tuple]` If we already have information about this tuple ensure that the ownership details are up-to-date. Otherwise add the tuple to our queue, satisfy any matching RD requests made on the local host (transmitting an ACCESS message for each one), then look for a matching IN request. If we find one, check whether we are the current owner, transmitting a DELETE or CHOWN_REQ as appropriate
IN	`[client_id, request_id, type, spec]` Should we have a matching tuple, multicast an appropriate OUT message, otherwise add the IN request to our queue
RD	`[type, spec]` Check if we have a matching tuple and if so multicast an OUT message
CHOWN_REQ	`[tuple_id, client_id]` First, check to see if we known about this tuple. If we don't, transmit a REPAIR_REQ. Should we know the tuple has been deleted, multicast a DELETE. If we own the tuple, we can transmit a CHOWN_ACK nominating the originator of the CHOWN_REQ as the new owner, otherwise we send a CHOWN_ACK stating who we understand to be the current owner
CHOWN_ACK	`[tuple_id, owner_id]` If we know about this tuple update its ownership. If we are the new owner and have a pending local IN which matches, service the request and multicast a DELETE message
DELETE	`[tuple_id, request_id]` Mark the unique tuple id as having been deleted and ensure both the tuple and the IN request it satisfied are removed from our cache
ACCESS	`[tuple_id]` If we don't know about this tuple, transmit a REPAIR_REQ, otherwise if we know it to have been deleted, multicast a DELETE message
REPAIR_REQ	`[tuple_id]` If we have this tuple multicast a REPAIR_ACK
REPAIR_ACK	`[tuple_id, owner_id, type, tuple]` Queue any unknown tuples

to a particular tuple space. Each tuple has a host-unique sequence number which can be used to detect missing tuples. When a tuple is unaccounted for, a repair message (REPAIR_REQ) is issued to request the retransmission of missing tuples and thus move closer to eventual global consistency. The structure of the network and the repair transmission backoff strategy implies that the 'nearest' platform who has the tuple cached will respond (with a REPAIR_ACK) first. If a host snoops an identical REPAIR_ACK message from another host, it avoids transmitting a response itself thus preventing acknowledgement implosion.

To speed up detection of missing tuples we use the ACCESS and DELETE messages when tuples are rd'd or in'd, respectively. These messages can be considered of comparatively low priority since they are used primarily to allow other hosts to detect missing tuples or prevent the use of stale tuples in rd requests. The earlier these messages are transmitted, the faster the independent views of a tuple space converge. However, as their delay does not alter the semantics of the tuple space, we can batch ACCESS's and DELETE's with other protocol traffic to reduce overall communication overhead.

The protocol messages, their format and usage are more comprehensively explained in Table 1 (first published in [12]).

The protocol has been designed with a degree of duplication to allow the protocol to recover from missed messages. For instance, when an application performs an `in` operation, the message corresponding to the operation will get propagated only if it cannot be satisfied from cached data. Therefore, in the case where the tuple data has been missed, the operation does not have to block until it is repaired. The redundancy in the protocol together with the local repair consistency mechanism allows hosts to connect and disconnect from tuple spaces at will and will ensure that consistency is eventually achieved. However, when a host 'reconnects' to a tuple space it can generate a flurry of repair messages. In the case of a low-bandwidth mobile network, this repair traffic is undesirable, particularly when the mobile host's client applications are only interested in particular subset of the available tuples. For this reason we have developed a simple proxy architecture which operates a split-level caching algorithm, allowing effectively redundant traffic to be filtered over low-bandwidth links (see Section 5).

3.4. Status

The current version of the platform consists of approximately 4000 lines of C for the daemon process and 500 lines for the API stubs. Bindings exist for the API in C, C++ and Java. We have used the platform to build a number of applications including a collaborative geographical information system (GIS), a group coordination service for a low bit rate video surveillance tool and a collaborative virtual environment.

3.4.1. Applications

Collaborative GIS. The collaborative GIS application allows a group of users to share spatial information using a shared whiteboard metaphor (see figure 2). A common map can be displayed, over which highlighting operations, such as placing text or drawing one of a variety of simple shapes, may be performed.

Any operation performed on one whiteboard is mirrored in all others. Each drawing operation is translated into a unique tuple which is propagated to a shared tuple space (the universal tuple space). Clients utilise the platform to discover new drawing operations (tuples). The distributed systems platform allows simple applications such as the GIS to share group state without complex group management or floor control protocols. Since tuples are persistent, all group state remains in the universal tuple space. This allows late entry into any GIS based collaboration and, furthermore, enables participants in the group activity to leave and later rejoin the collaboration at any point.

Video application. Based on the requirements of the Emergency Services [34] we have developed a rapidly deployable camera server demonstrator (see figure 3). This application

Figure 2. The collaborative GIS tool.

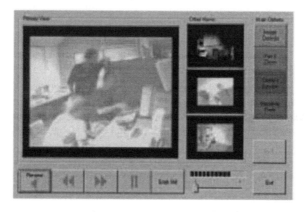

Figure 3. The rapidly deployable camera server demonstrator.

allows users to quickly set-up ad-hoc surveillance systems and view the resulting video streams on mobile devices.

Each surveillance unit is capable of capturing frames of digital video, compressing them using a low-bitrate H.263 CODEC and transmitting them over a wireless local area network (in this case WaveLAN). A backchannel from the client allows the user to control the encoding process at the video source: the client may adjust the frame size and various camera parameters including contrast, brightness and so on.

As new camera services become available a tuple is output describing the camera, it's field of view and position together with a communications end-point. The application removes the tuple from the tuple space and establishes a connection to build a low-resolution thumbnail view. The user of the application may choose to enhance the view if the output from the camera is of particular interest. The tuple space allows the demonstrator to be highly flexible, we can deploy both cameras and the application at any point of our network and the demonstrator will dynamically reconfigure itself.

Virtual collaborative environments.   Recent years have witnessed the use of spatial meta-phors for presenting multimedia information. This has been recognised in the Inhabited In-formation Spaces schema of the I^3 initiative. The eSCAPE project is seeking to develop more inclusive virtual environments that encompass and involve the citizen. More specifically, eSCAPE is developing two thematic places that act as contrasting alternative instantiations of electronic landscapes.

In the first of these places, the electronic realisation of where points meet in virtual space will be instantiated as a Virtual Cityscape. The Cityscape provides a concrete and familiar metaphor as a means of visualising the interconnection between shared virtual environments. The second thematic representation will represent the interconnection of virtual places as a Virtual Planetarium where shared virtual environments and the connections between them can be represented in more abstract forms.

The L^2imbo platform is being used as a communications substrate to provide distributed sharing of components of these virtual worlds. These components, which are built using a toolkit consisting of Java Beans, communicate by building tuple representations of them-selves which are then distributed using L^2imbo. Tuples are thereby used to represent both the real world data objects being modelled and their geometric virtual counterparts.

3.4.2. Performance.   To evaluate the relative performance of L^2imbo with respect to ex-isting distributed systems platforms, we have compared it against both the ANSAware distributed systems platform (version 4.1) [2] and raw BSD sockets. Our test suite con-sisted of three separate pairs of client and server processes which carry out 1000 timed RPC interactions consisting of an *n* byte payload and null response. The test configuration was a pair of SparcStation 1 workstations networked with a moderately loaded 10 Mbps Ethernet.

An RPC is modelled in the tuple space by the exchange of two tuples with types *request* and *response*, respectively. It is important to note that a directed communication of this kind is not well suited to the tuple space paradigm, as it represents two tightly couple tuple insertions and removals. However, we believe that it is important to demonstrate that the worst case performance of our platform is still within acceptable bounds.

To isolate the additional overhead of splitting the L^2imbo platform into separate processes (daemon and client libraries), we have run tests for both an optimised form (in which the daemon and API processes are linked into a single executable) and unoptimised (separate processes) forms of the client and server. The results are summarised in Table 2.

The figures show that in optimised trim, L^2imbo outperforms ANSAware RPC in all cases. However, the overhead of the context switch and local communication required in the standard L^2imbo configuration does have a significant impact on the figures. Being able to minimise the overheads associated with exchanging messages between the application stubs and the daemon process is clearly an important factor in further improving the performance of L^2imbo.

Table 2.   Comparison of relative performance on SunOS.

Payload (bytes)	Sockets (UDP)	ANSAware 4.1 (REX)	L^{2}imbo DTS (linked)	L^{2}imbo DTS (separate processes)
256	2.98	7.10	6.53	12.58
512	3.45	10.48	7.20	13.47
1024	3.93	11.17	8.64	15.10
2048	5.85	13.14	11.97	20.28
4096	9.46	21.14	18.06	28.26
8192	15.83	34.83	29.93	44.82

We also compared our results with published figures for the Chorus Systèmes COOL ORB [7] on the Linux platform. The figures suggest that the performance of L^{2}imbo is comparable to conventional RPC based distributed systems platforms in RPC-like tests. More specifically, the COOL benchmark report quotes 3.8 ms for a basic request exchange of 1000 bytes in each direction. On a similar specification Linux platform, the linked version of L^{2}imbo takes 4.4 ms to perform this same test (averaged over 1000 interactions). Furthermore, for interactions of 100 bytes in each direction, COOL is quoted as taking 2.6 ms, whereas the optimised form of L^{2}imbo takes just 1.9 ms. Note that we have not attempted to verify the published figures for COOL.

In considering these figures it is important to note that the test case demonstrates directed communication. In the conventional distributed systems platform the timing information is taken *after* an initial process of binding and thus represents the best possible case for these platforms. In the case of L^{2}imbo however, the test case represents a worst case scenario; the tuples are being rapidly inserted and removed from the tuple space and the overhead associated with matching is not strictly necessary since only two well known processes are communicating.

4.   Application programmer interface issues

One of the most appealing features of the tuple space model is the simple yet expressive API. As discussed in Section 3.1, we have implemented this fundamental set of operations (`in`, `out`, `rd` and `eval`) in the L^{2}imbo platform.

While it is true to say that one can construct any arbitrary application using the basic Linda API, we believe that performance and usability (particularly in distributed platforms) can be improved through the use of some additional, carefully chosen, operations.

4.1.   Enumeration

Consider the case where an application was required to count the number of tuples matching a certain criteria. Ideally, such an application would simply access each tuple in turn, without removing each of them from the tuple space, using the `rd` operation. Using the standard

API calls however, the application could not guarantee that each successive rd call would not return the same tuple over and over again, since both in and rd operations match tuples non-deterministically (this scenario is, as briefly discussed in Section 2.3.1, known as the multiple-rd problem). The application would therefore have to remove each of the matching tuples using the in operation one at a time and, only when all of the relevant tuples were removed, would have to insert them back into the tuple space again. However, there is a further problem. As the application can not know how many tuples are going to match its criteria, it cannot know how many tuples to read in and must take some measure to ensure it doesn't block indefinitely. In short, using the standard API the application or applications generating the tuples would have to be restructured.

To address these problems, we have implemented four further operators: inp and rdp [27], collect and copy-collect [31]. The first two of these are non-blocking versions of in and rd which evaluate to boolean values indicating their success. This allows application programmers to avoid having to determine for themselves when a blocking in or rd operation is not going to return.

The collect and copy-collect operations are designed to optimise situations where tuple enumeration is required. Collect yields all the tuples matching a particular anti-tuple, copying each of them to a particular destination tuple space. Collect returns the number of tuples it copied to the API allowing the application to enumerate through the results more easily. The copy-collect operator behaves identically to collect except it does not remove the matching tuples. The two operations are analogous to in and rd, respectively.

4.2. Asynchronous operation

We have implemented a superset of Rowstron's Bonita primitives [31]. These enhancements enable tuple space clients to access tuple spaces asynchronously by dispatching requests using one primitive and later collecting the results using another. A further operator allows clients to poll the platform to determine whether the results for a previous request are available.

The Bonita primitives have two primary uses. Firstly, the primitives allow applications to be structured with less internal parallelism, since multiple operations can be executed simultaneously from a single thread of execution. The second reason is one of performance enhancement; the asynchronous primitives are analogous to splitting the standard synchronous in and rd operations into two phases; a request and yielding the results. In the situation where an application needs only to determine if a certain matching tuple exists and does not need the specific actuals from the matching tuple, then it is more efficient to use the asynchronous operation as the results do not need to be passed to the application and unmarshalled into application variables.

Note that while Rowstron proposes an overloaded dispatch primitive to issue asynchronous in, out and rd requests, we instead use separate out, async_in and async_rd for enhanced code clarity.

4.3. Eventing services

One of the more interesting applications of the L^{2}imbo platform has been in supporting collaborative working in distributed virtual worlds under the auspices of the eSCAPE project.

In the eSCAPE architecture, multimedia information such as documents and video clips can be exchanged using the tuple space. Furthermore, the platform is also used at a fundamental level to facilitate sharing of the actual geometric information that comprises the virtual world itself.

This work has highlighted several issues concerning using the tuple space as a medium for supporting the sharing of timely information. Due to the non-deterministic matching of tuples with anti-tuples, it is difficult for applications to efficiently determine when new tuples of a certain type arrive. For example, if an application is producing a tuple describing the coordinates of a moving object in a virtual world, then other collaborating applications need to be able to identify the most recent set of coordinates. In existing tuple space platforms this functionality could only be achieved by layering timing or sequencing information in the tuples themselves, then enumerating through all the available tuples to find the most recent.

In L^2imbo we have chosen to solve this problem by adding simple eventing properties to the platform. The register operation allows applications to receive notification callbacks when new tuples matching a particular anti-tuple is observed in the tuple space. The register operation is functionally equivalent to a temporally extended `rd` operation (which can not yield the same tuple more than once). Register is functionality akin to the `notify` operation of Sun's JavaSpaces platform [21].

4.4. Analysis

The L^2imbo platform API has evolved as it has been used to support more diverse application domains. The full platform API (as of version 1.3) is shown in Table 3.

Experience has shown that the most demanding application of the tuple space has been to support peer-to-peer collaborative activity. This application domain is far removed from the Gelernter's original tuple space work. However, it is important to note that we have changed only the API and not the paradigm to support these new domains.

What the platform has gained in functionality however, it has lost in simplicity and elegance. We are therefore currently aiming to distil the concepts from the current API into a more elegant set of operations. We have identified the following issues that we will need to address with any new API:

1. The API must be expressive enough to allow applications to enumerate through those tuples that match a given anti-tuple efficiently. Moreover, most applications require more flexible matching than the simple equality tests supported by the current API.
2. An application must be able to find out whether a particular tuple is currently contained in the tuple space simply and quickly, without requiring complex application semantic changes.
3. Some applications would be easier to engineer if more complex data types were allowed within tuple fields (e.g., lists, sets and so on). New data types imply new matching criteria, such as maximise, minimise and contained within relations. Such relations are not possible in any existing tuple space API and do not obviously fit in most programming languages.
4. Lastly, the tuples in current tuple space paradigms are persistent and will remain until explicitly removed. Some applications seem to require new classes of tuples which are

Table 3. The platform API.

Primitive	Syntax and description
use	`ts = use(handle)` Ensures the local tuple space daemon is operating a local cache for the specified unique tuple space handle and returns a local identifier for this tuple space
discard	`discard(ts)` Used by clients to inform the local daemon that they no longer require access to the specified tuple space
out	`out(ts,type,<ACTUAL variable|UNDECLARED>,...)` The standard Linda `out` primitive
async_in	`rqid = async_in(ts,type,<ACTUAL variable|FORMAL value>,...)` A non-blocking operation used to withdraw a matching tuple from tuple space. `async_in` dispatches a request to the L^2imbo platform and returns a request identifier to the client. A matching reply can later be retrieved by the client using `obtain` (described below) with this identifier
async_rd	`rqid = async_rd(ts,type,<ACTUAL variable|FORMAL value>,...)` As `async_in`, expect that matching tuples are not withdrawn from tuple space
arrived	`boolean = arrived(rqid)` A non-blocking primitive which returns a boolean indicating whether a tuple satisfying request `rqid` is available from the client stub
obtain	`obtain(rqid)` Blocks the client until a tuple satisfying request `rqid` is received by the client stub. When a result is available, `obtain` populates any formal variables passed in the anti-tuple of the `async_in` or `async_rd` operation which created `rqid` with the appropriate values from the matched tuple
cancel	`cancel(rqid)` Used by clients to cancel their need for a response to a previous `async_in` or `async_rd` request
in	`in(ts,type,<ACTUAL variable|FORMAL value>,...)` The standard Linda `in` primitive
rd	`rd(ts,type,<ACTUAL variable|FORMAL value>,...)` The standard Linda `rd` primitive
collect	`collect(srcid,dstid,type,<ACTUAL variable|FORMAL value>,...)` Collect all tuples in the source tuple space which match the specified anti-tuple and move them to the destination tuple space
copy_collect	`copy_collect(srcid,dstid,type,<ACTUAL variable|FORMAL value>,...)` Find all tuples in the source tuple space which match the specified anti-tuple and make a copy of them in the destination tuple space
register	`register ts,ARRIVAL|DELETION, type, callback)` Register for notification of the arrival/deletion of a particular tuple type
deregister	`deregister(tsId)` Deregister for event notification

'transient', i.e., contain information which is valid only for a particular period of time. The inclusion of such tuples require an adjustment to the tuple space paradigm and have an impact in both engineering and API terms.

Based on our experiences of developing and using tuple space based platforms, we believe the above issues represent serious concerns for researchers considering using this paradigm. More specifically, before a tuple space platform can become widely applicable, solutions will need to be found to these problems.

5. Adaptation

The concept of adaptation is central to a modern distributed systems platform. In particular, platforms which wish to support either multimedia data or mobile operation must be able to adapt their behaviour in response to changes in their environment. Adaptation can be divided into a number of distinct facets, i.e., QoS and context monitoring, reporting and adaptation. It is important to stress at this point that we are not concerned with adaptation solely in response to changes in communications QoS but rather as a consequence of changes in a wide range of environmental factors including communications QoS and cost, physical location, power and device availability and user preferences.

Our basic approach is to use a local tuple space to support QoS reporting (see figure 4). More specifically, we allow multiple, self-contained, QoS monitoring agents to deposit information into a local tuple space. This information takes the form of tuples of arbitrary types. Applications and system components which are interested in receiving this QoS information are able to carry out a rd operation on the local tuple space, thus non-destructively obtaining a copy of the information.

There are a number of significant advantages to this approach over conventional event based QoS architectures as typified by [8]. Firstly, the local tuple space acts as an effective bi-directional 'layer-breaker'. This enables applications to both monitor and control

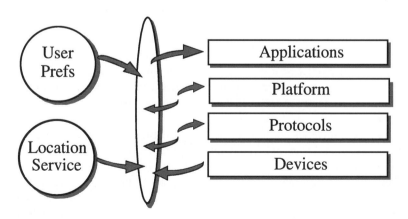

Figure 4.   A local QoS tuple space.

underlying system components without requiring existing APIs to be modified. Secondly, by using a tuple space we are able to achieve temporal de-coupling between the producers and consumers of QoS information. This enables, for example, QoS monitors to run as periodic processes which deposit information into the QoS tuple space for consumption by applications some arbitrary time in the future. Examples of such monitors include position sensors which sample a local GPS compass and report on the user's physical position and communications monitors which are able to report on the state of the end-system's network interfaces. Finally, since QoS information is placed in the QoS tuple space as tuples they may be propagated (possibly using bridging agents) to other applications and end-systems. This enables nodes to determine the QoS of other nodes without requiring explicit application level support.

However, despite the cited advantages of our approach the tuple space paradigm presents a number of unique challenges with respect to the support of adaptation. These challenges are discussed in detail in the following sections.

5.1. Communications QoS monitoring

Conventional QoS monitoring techniques rely on the establishment of end-to-end communications paths. Traffic patterns on these paths can then be measured and reported to applications and users to enable adaptation. For example, the MOST platform used a modified version of the REX RPC package to monitor round-trip times between application level objects. These times were mapped onto estimations of latency and bandwidth and reported back to the applications involved in the communications. Using this technique programmers could construct applications which automatically adjusted their demands on the underlying communications infrastructure in line with changes in resource availability. Similar approaches have been used in a range of mobile platforms including [23].

However, in a tuple space platform the concept of an end-to-end communications path is lost with the removal of the concept of directed communications and bindings. As a consequence, the production of communications related QoS figures becomes problematic. Furthermore, the production of such figures looses much of its significance since in many cases the eventual consumer of a tuple is not known in advance. Hence, it is not possible to provide predictions of, for example, the time taken to deliver a tuple to its destination (not least because the destination will depend on the existence of matching tuple requests on different nodes). Indeed, if we consider a scenario in which an end-system is disconnected then tuples issued by an application may be matched by a local service in which case propagation will be nearly instantaneous or, may not be propagated until connectivity is restored.

In spite of the difficulties outlined above, it is possible to provide an indication of connectivity on a per-node basis. In L^2imbo this can be achieved using QoS monitoring agents which observe events at various points in the system including the: rate of injection of tuples into a given tuple space, rate of access to tuples in tuple space (through `in` or `rd` operations), the cost of the current channel and the level of connectivity (in terms of the raw capabilities of the network interface). The results of the QoS monitoring can be placed in the local QoS tuple space and, if required, propagated to other nodes (see previous section).

This information can be used to, for example, start up local proxies to service application requests or to adapt application behaviour.

Finally, we observe that if end-to-end QoS is important this can be engineered using tuple spaces with restricted memberships and additional QoS features such as admission control and traffic monitoring. An architecture for such a system is described in [4]. However, such an approach negates many of the benefits of using tuple spaces including temporal de-coupling and undirected communications.

5.2. *Adaptation*

Given the problems in monitoring QoS in a tuple space based systems adaptation is often carried out on, or by, individual end-systems in response to changes in the availability of local network interfaces. More specifically, when an end-system is experiencing poor network connectivity it may take steps to optimise its use of available network resources. The simplest form of adaptation is for the end-system to instantiate local proxy services to deal with application requests. Such services are able to interact with applications without the need for additional support since communications is, in any case, undirected.

A more complex form of adaptation involving the use of proxies is, however, possible. Proxies are the subject of intense research in the mobile computing community [16, 32, 35, 36]. In such systems a proxy is usually defined as an entity inserted into a communication stream somewhere between the client and server (see figure 5). This proxy can then modify the communication streams (either from client to server or vice versa) in order to adapt this technique to the current QoS characteristics. This adaptation can be done using one or more of the following basic proxy actions:

- *Filtering*: A filtering proxy can reduce the amount of data to be transmitted by filtering out less relevant data making it loss-prone. An example could be the omission of colour information from a video stream.
- *Transforming*: A transforming proxy can change the data type of the stream it processes to either reduce the data volume or to adapt the data format to be easily presentable on

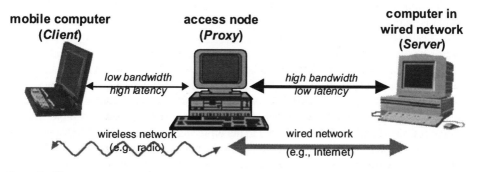

Figure 5.   The concept of a proxy.

the client. For example, the proxy could transform a given PostScript document into a plain ASCII file, preserving the textual contents but omitting formatting information and graphics.

- *Caching*: Finally, a caching proxy can be used to add asynchronicity to a communication stream decoupling it from bad transmission characteristics. Several caching strategies can be implemented by a proxy: An on-demand caching proxy could be used to minimise the amount of data to transmit over a link with bad QoS, while a pre-fetching proxy could utilise the link if the current QoS is good.

The above functions can, of course, be easily combined to allow, for example, filtering and transforming proxies. Since proxies rely on being inserted into the communications path between client and server it may appear at first glance that they have little relevance for tuple space platforms. However, if we view end-systems as having a point of connection to a tuple space this forms a natural point for installing proxies. In the L^{2}imbo architecture we support this idea using a number of additional system services:

(i) *Tunnel agents*: Tunnel agents are used to connect to a tuple space on an end-system's behalf and to propagate (a subset of) tuples and tuple requests to the end-system. For example, if an end-system was accessing a tuple space via a GSM link a tunnel agent could be used to receive the tuples on the fixed network and to forward them to the mobile host. Note that tunnels are distinct from the bridging agents described in Section 3 since they are intra rather than inter tuple space. Tunnel agents can be configured to only forward tuples of a specific type or which match a certain anti-tuple. This should be viewed as a delay in the transmission of certain classes of tuples rather than a discard operation since the tuples will remain available in the tuple space.

(ii) *Filter agents*: Filter agents `rd` tuples from the tuple space and place new versions of the tuples back into the tuple space. For example, a generic compression filter agent may read in all tuples as they arrive, compress the payload and output the compressed version as a new tuple.

Such an architecture has a number of key advantages over a conventional proxy architecture. Firstly, it maintains a clean separation between the selective transmission of traffic, i.e., deciding which tuples to propagate, and the conversion of traffic (tuples). This separation is important because it enables the two concerns to be addressed independently. More specifically, it would be perfectly possible to set up a tunnel agent to delay tuples of a specific type and to use a third party filter agent to convert tuples from this type into a more acceptable format which can be transmitted straight away. Secondly, the results of a tuple conversion are, conceptually at least, propagated to other members of the tuple space. As a result, a single filter agent can carry out conversions which are of use to many tunnel agents. Finally, since tuples persist it is possible for clients who are connected to a tuple space to receive converted tuples while connected to the tuple space by a low-speed link and to then retrieve the original versions of the tuples when they reconnect using a high-speed link.

Despite many promising aspects the above architecture requires significant further work before we can claim to have addressed the issues of proxies in tuple space based platforms.

6. Concluding remarks

At first glance the tuple space paradigm appears to offer solutions to many of the problems inherent in mobile computing. In particular, the temporal de-coupling allows applications to survive periods of disconnection and the use of anonymous, undirected communications allows transparent rebinding to proxy services. Furthermore, the apparent ease with which group applications may be supported makes the paradigm appear attractive for use in collaborative applications running on both fixed and wireless networks. Motivated by these perceived benefits the authors have developed a comprehensive tuple space based platform called L^2imbo. The platform offers a comprehensive selection of features including multiple tuple spaces, tuple typing and asynchronous operations. These features are further extended by the addition of a range of system services which provide facilities for tuple management and QoS monitoring and control. Our platform has been ported to SunOS, Solaris, Linux, Windows 95 and Windows NT and supports C, C++ and Java APIs. It has been used to support a number of applications including collaborative editors and virtual environments.

The development of this platform and its subsequent use to support a range of mobile applications has provided us with a valuable insight into the practical problems of using the tuple space paradigm to support mobile applications.

These problems can be divided into two distinct areas, the API and system support for adaptation. With respect to the API the problems focus on the suitability of the basic tuple space API for constructing general distributed applications. In our experience, programmers have substantial difficulties developing applications which use only the basic Linda API. More significantly, the addition of new features such as asynchronous operators and solutions to the multiple-rd problem only partially address these problems. In particular, we believe developers of collaborative applications require generalised event support and all application developers require significantly extended tuple data types and matching capabilities. However, adding such features to a platform such as L^2imbo effectively transforms the platform into a distributed database with support for events.

Supporting adaptation in a tuple space based platform also causes serious problems. Most significantly, the lack of an explicit representation of the end-to-end communications path (i.e., a binding) makes it difficult to monitor and adapt to fluctuations in QoS. This represents a fundamental dichotomy for developers of tuple space platforms. The absence of bindings brings significant advantages in terms of transparent service rebinding etc. but makes it impossible to develop adaptive applications. As a consequence we believe that tuple spaces represent only part of the solution: for many applications and communications scenarios explicit bindings with QoS will be required.

Despite the comments we have made above, there are a number of positive features of the tuple space paradigm. In particular, the use of a tuple space as a means of disseminating QoS information appears extremely promising. More specifically, the undirected communications and temporal de-coupling appear ideally suited to the propagation of information between applications and system components on the local host (the ability to propagate this information using bridging agents and a distributed tuple space, while initially attractive, requires significant further research since the lack of propagation guarantees inherent in tuple space platforms may impinge on the general usefulness of such an approach). As a consequence we are currently investigating the use of tuple spaces as a component within a more generalised distributed systems framework. More specifically, we are considering the addition of tuple spaces to CORBA like platforms. We anticipate, in the first instance, that local tuple spaces will be used for QoS propagation. The integration of distributed tuple spaces into such an architecture raises significant new research issues in terms of the use of multiple communications paradigms within a single application framework.

Acknowledgments

This work was carried out under the auspices of the EPSRC funded Reactive Services project.

Note

1. It is possible to produce tuples for an identified consumer by encapsulating destination information in tuples. This is termed *directed communications*. Several schemes to achieve this have been proposed including an approach based on Amoeba ports [29].

References

1. E. Amir, H. Balakrishnan, S. Seshan, and R. Katz, "Efficient TCP over networks with wireless links," in Proc. 5th IEEE Workshop on Hot Topics in Operating Systems (HotOS-V), Rosario Resort, Orcas Island, Washington, US, IEEE Computer Society Press, May 1995.
2. A.P.M. Ltd., "ANSA: An engineer's introduction to the architecture," Architecture Projects Management Ltd., Cambridge, UK, November 1989.
3. A. Bakre and B.R. Badrinath, "M-RPC: A remote procedure call service for mobile clients," Technical Report WINLAB TR-98, Department of Computer Science, Rutgers University, US, June 1995.
4. G.S. Blair, N. Davies, A. Friday, and S.P. Wade, "Quality of service support in a mobile environment: An approach based on tuple spaces," in Proceedings of the 5th IFIP International Workshop on Quality of Service (IWQoS '97)—Building QoS into Distributed Systems, Columbia University, New York, US, May 1997, pp. 37–48.
5. N. Carriero, D. Gelernter, and L. Zuck, Bauhaus Linda, Selected Papers from ECOOP '94, Bologna, Italy, July 1994, pp. 66–76.
6. K. Cheverst, N. Davies, A. Friday, and G.S. Blair, "Services to support consistency in mobile collaborative applications," in Proc. 3rd International Workshop on Services in Distributed Networked Environments (SDNE), Macau, China, IEEE Computer Society Press, June 1996, pp. 27–34.
7. Chorus Systèmes, "CHORUS/COOL-ORB programmer's guide," Technical Report CS/TR-96-2.1, Chorus Systèmes, 1996.
8. G. Coulson, G.S. Blair, F. Horn, L. Hazard, and J.B. Stefani, "Supporting the real-time requirements of continuous media in open distributed processing," Computer Networks and ISDN Systems, 1994, to appear.
9. N. Davies, G. Blair, K. Cheverst, and A. Friday, "Supporting adaptive services in a heterogeneous mobile environment," In Proc. Workshop on Mobile Computing Systems and Applications (MCSA), Santa Cruz, CA, US, Luis-Felipe Cabrera and Mahadev Satyanarayanan (Eds.), IEEE Computer Society Press, December 1994, pp. 153–157.
10. N. Davies, G.S. Blair, K. Cheverst, and A. Friday, "Supporting collaborative applications in a heterogeneous mobile environment," Special Issue of Computer Communications on Mobile Computing, vol. 19, pp. 346–358, 1995.

11. N. Davies, A. Friday, S. Wade, and G. Blair, "L²imbo: A distributed systems platform for mobile computing," ACM Mobile Networks and Applications (MONET), Special Issue on Protocols and Software Paradigms of Mobile Networks, vol. 3, no. 2, pp. 143–156, 1998.

12. N. Davies, A. Friday, S. Wade, and G. Blair, "An asynchronous distributed systems platform for heterogeneous environments," in Proc. 8th ACM SIGOPS European Workshop: Support for Composing Distributed Applications, Sintra, Portugal, ACM Press, 1998.

13. N. Davies, S.P. Wade, A. Friday, and G.S. Blair, "Limbo: A tuple space based platform for adaptive mobile applications," In Proceedings of the International Conference on Open Distributed Processing/Distributed Platforms (ICODP/ICDP '97), Toronto, Canada, May 1997, pp. 291–302.

14. A. Douglas, A. Wood, and A. Rowstron, "Linda implementation revisited," Transputer and Occam Developments, IOS Press, 1995, pp. 125–138.

15. S. Floyd, V. Jacobson, S. McCanne, C. Liu, and L. Zhang, "A reliable multicast framework for light-weight sessions and application level framing," in Proceedings of ACM SIGCOMM '95, Cambridge, Massachusetts, US, ACM Press, August 1995, pp. 342–356.

16. A. Fox, S.D. Gribble, E.A. Brewer, and E. Amir, "Adapting to network and client variation via on-demand, dynamic distillation," in Proc. ASPLOS-VII, Boston, MA, US.

17. A. Friday, G.S. Blair, K.W.J. Cheverst, and N. Davies, "Extensions to ANSAware for advanced mobile applications," in Proc. International Conference on Distributed Platforms, Dresden, A. Schill, C. Mittasch, and O. Spaniol (Eds.), Chapman and Hall, pp. 29–43.

18. D. Gelernter, "Generative communication in Linda," ACM Transactions on Programming Languages and Systems, vol. 7, no. 1, pp. 80–112, 1985.

19. D. Gelernter, N. Carriero, S. Chandran, and S. Chang, "Parallel programming in Linda," in Proceedings of the International Conference on Parallel Processing, August 1985, pp. 255–263.

20. B. Grönvall, I. Marsh, and S. Pink, "A multicast-based distributed file system for the internet," in Proceedings of the 7th ACM SIGOPS European Workshop, Connemara, Ireland, ACM Press, September 1996.

21. "Sun's JavaSpaces is foundation for future distributed systems," SunWorld, August 1997.

22. S. Hupfer, "Melinda: Linda with multiple tuple spaces," Technical Report YALEU/DCS/RR-766, Department of Computer Science, Yale University, New Haven, Connecticut, US, February 1990.

23. A. Joseph, A. deLespinasse, J. Tauber, D. Gifford, and M.F. Kaashoek, "Rover: A toolkit for mobile information access," in Proc. 15th ACM Symposium on Operating System Principles (SOSP), Copper Mountain Resort, Colorado, US, ACM Press, vol. 29, December 1995, pp. 156–171.

24. A.D. Joseph and M.F. Kaashoek, "Building reliable mobile-aware applications using the rover toolkit," Technical Report, M.I.T. Laboratory for Computer Science, 1996.

25. R.H. Katz, "Adaptation and mobility in wireless information systems," IEEE Personal Communications, vol. 1, no. 1, pp. 6–17, 1994.

26. R. Katz and E. Brewer, "The case for wireless overlay networks," in Proc. SPIE Multimedia and Networking Conference (MMNC), San Jose, CA, US, January 1996.

27. J.S. Leichter, "Shared tuple memories, shared memories, buses and LAN's—Linda Implementations across the Spectrum of Connectivity," Ph.D. Thesis, Department of Computer Science, Yale University, New Haven, Connecticut, US July 1989.

28. J. Nieh and M. Lam, "Integrated processor scheduling for multimedia," in Proc. 5th International Workshop on Network and Operating System Support for Digital Audio and Video (NOSSDAV), Durham, New Hampshire, US, April 1995.

29. J. Pinakis, "The design and implementation of a distributed Linda tuple space," in Proceedings of the 2nd Department of Computer Science Research Conference, Department of Computer Science, University of Western Australia, Nedlands, WA 6009, 1991.

30. A. Rowstron and A. Wood, "Solving the Linda multiple rd problem," in Proc. Coordination Languages and Models (Coordination '96).

31. A.I.T. Rowstron and A.M. Wood, "Bonita: A set of tuple space primitives for distributed coordinartion," in Proceedings of the 30th Annual Hawaii International Conference on System Sciences, IEEE CS Press, vol. 1, 1997, pp. 379–388.

32. J. Seitz, N. Davies, M. Ebner, and A. Friday, "A CORBA-based proxy architecture for mobile multimedia applications," in Proc. 2nd IFIP/IEEE International Conference on Management of Multimedia Networks and Services (MMNS '98), Versailles, France.

33. A. Xu and B. Liskov, "A design for a fault-tolerant, distributed implementation of Linda," in Proceedings of the 19th International Symposium on Fault-Tolerant Computing, June 1989, pp. 199–206.
34. N. Yeadon, N. Davies, A. Friday, and G.S. Blair, "Supporting video in heterogeneous environments," In Proc. Symposium on Applied Computing, Atlanta, US.
35. N. Yeadon, F. Garcia, D. Hutchison, and D. Shepherd, "Filters: QoS support mechanisms for multipeer communications," Journal on Selected Areas in Communications, JSAC, vol. 14, no. 7, pp. 1245–1262, 1996.
36. B. Zenel and D. Duchamp, "Intelligent communication filtering for limited bandwidth environments," in Proc. 5th IEEE Workshop on Hot Topics in Operating Systems (HotOS-V), Rosario Resort, Orcas Island, Washington, US, IEEE Computer Society Press, May 1995.

Distributed and Parallel Databases, 7, 343–372 (1999)
© 1999 Kluwer Academic Publishers. Manufactured in The Netherlands.

MODEC: A Multi-Granularity Mobile Object-Oriented Database Caching Mechanism, Prototype and Performance

BORIS Y.L. CHAN   csylchan@comp.polyu.edu.hk
HONG VA LEONG   cshleong@comp.polyu.edu.hk
Department of Computing, The Hong Kong Polytechnic University, Hung Hom, Hong Kong

ANTONIO SI   antonio.si@eng.sun.com
Sun Microsystems Inc., Palo Alto, CA 94303, USA

KAM FAI WONG   kfwong@se.cuhk.edu.hk
Department of Systems Engineering and Engineering Management, The Chinese University of Hong Kong, Shatin, Hong Kong

Received October 1, 1998; Accepted

Recommended by:

Abstract. An inherent limitation in mobile data access is due to the unreliable and low bandwidth wireless communication channel. Caching of useful database items from database server in local storage of mobile clients is effective in reducing data access latency and wireless bandwidth consumption. In the event of disconnection, cached data can also serve the purpose of partial query processing. In this paper, we present the implementation and evaluate a new caching mechanism for object-oriented database systems in a mobile environment called MODEC. MODEC possesses the capabilities of performing caching at multiple granularities and adapting to changes in data access pattern, providing improved performance through tolerating limited inconsistency to read-only transactions. This caching capabilities is supported via standard ODMG modeling constructs. The prototype of MODEC is implemented using ODE database. Empirical system performance results are obtained from experiments on the prototype with data from a real-life database. The results are validated against results obtained via detailed simulation studies on MODEC. Both sets of results are found to be consistent and are in favor of our MODEC mechanism in providing a feasible solution to the mobile data access problem under the constraints in a mobile environment.

Keywords: mobile data access, distributed database, multi-granularity caching, cache replacement, cache refreshing

1.   Introduction

Mobile environment has become prominent as advances in wireless communication technology accelerate during recent years. The penetration rate of mobile phones and related products has increased at an unanticipated rate. In 1998, Hong Kong and the United States rank fourth and fifth, respectively, in the world in terms of mobile penetration rate, just behind the Scandinavian countries. Access to stationary database server from mobile clients has also become a norm rather than an exception. Data transfer occurs on a wireless communication channel with a rather low bandwidth. Usually, a wireless channel has an uplink bandwidth of 19.2 Kbps and a downlink bandwidth up to a maximum of 1 or 2 Mbps, with a station of strong power. Mobile clients must compete for the scarce resource. Wireless channels also suffer from signal distortion, transmission error, and disconnection [4].

To reduce the reliance of a mobile client on the database server via vulnerable wireless channels, we have proposed the paradigm of *mobile caching* in [10]. This is known as Multi-granularity Mobile Object-oriented DatabasE Caching mechanism or MODEC. MODEC allows frequently accessed database items, often known as the *hot spot*, to be cached in the local storage of a mobile client. To combat for the low bandwidth and high transmission overhead, caching granularity can be varied from as fine as an attribute to as coarse as an object in the context of OODB. To determine the importance of a database item to be cached, a replacement score is estimated for the item from its past access history. The use of the score in selecting the replacement victim defines the cache replacement algorithm.

Another important issue in reusing cached database item is to determine the freshness of the item. The concept of Leases [19] has been adopted in file based systems to ensure the freshness of a remote file. A similar concept known as *refresh time* has been adopted in MODEC. The parameter that affects the performance and accuracy of the lease is the estimation function for the lease period. Too large a value will cause an out-dated database item to be reused without being noticed. Too small a value will cause too many unnecessary update requests initiated by a mobile client. We proposed an adaptive approach in estimating such a value and demonstrated its usefulness using two performance metrics: *error rate* and *false alarm*. We demonstrated that the adaptive approach using statistical properties of the updates occurring at the database server can yield better performance by delivering a better estimate for the refresh time.

In this paper, we conduct simulation studies on the proposed MODEC framework. We also describe a prototype for MODEC on an OODB called ODE (Object Database Environment) [2] from AT&T Laboratory. Empirical results are obtained from the MODEC prototype and the results are validated against the simulation results. The results show the improvement of data access. The consistency of both sets of results also reinforces our brief that MODEC is effective in a mobile database environment. We focus our discussion on a point-to-point communication paradigm between a mobile client and the database server in this paper. A similar multi-granularity caching mechanism based on the broadcast paradigm and the hybrid point-to-point and broadcast paradigm is left for future work.

This paper is organized as follows. In Section 2, a brief survey is conducted on related work, including caching and concurrency control issues. We give a brief review on our MODEC model in Section 3. Various issues of MODEC including cache granularity, cache replacement, cache coherence, and transaction processing are discussed. In Section 4, the implementation of the prototype is described. This is followed by a description on a series of conducted experiments and relevant empirical results in Section 5. A comparison and validation with simulation results are also presented. In addition to demonstrating that MODEC is useful in a mobile environment, the agreement we found between simulation and prototype results allows us to evaluate new protocols and algorithms more readily via the less time-consuming and flexible simulation studies, before being put forward into practice. Finally, we conclude this paper with a concluding remarks and highlight some possible future work in Section 6.

2. Related work

There are two major areas of consideration in a mobile caching environment: the management of individual database items in the cache and the consistency maintenance of a collection of database items as a whole. The former issue is related to traditional cache management and the latter issue is tied with transaction processing.

2.1. Cache management

In a client-server database environment, to combat the network transmission latency, a multi-level caching scheme could be established by caching database items from the server in a client's local memory and/or local storage. A storage cache also has an advantage of persistence. When disconnected from the server, a client can still operate on the database items available in its storage. A data caching scheme is usually characterized by three different issues: *cache granularity*, *cache coherence*, and *cache replacement*. We briefly survey each in this section.

2.1.1. Cache granularity. Caching mechanisms in conventional client-server environment are usually page-based [7, 17], primarily because the transmission bandwidth is comparable to the disk bandwidth and because the server's storage and operating system are also page-based. However, page-based transmission over wireless channels consumes a lot of bandwidth, as can be illustrated by the following simple example. Given a database system with a typical specification:

Total number of objects $(n) = 2000$

Size of each object (s) $= 128$ bytes

Page size (p) $= 4K$ bytes.

The number of pages (m) to store all objects is

$$m = 2000 \times 128/4\,\mathrm{K} = 62.5 \approx 63 \text{ (roundup)}.$$

According to Cardenas, the estimated number of pages needed to be accessed, X_D, in order to satisfy a query with a selectivity of k objects is defined by: $X_D = m(1 - (1 - 1/m)^k)$ [6]. With $k = 20$ objects, X_D would be 17.25 pages on average. To transfer 17.25 pages via a wireless channel of 19.2 Kbps requires 28.75 s, while it only takes 0.05 s to transfer these pages in an Ethernet with a bandwidth of 10 Mbps. It may not be worthy to employ page-based caching in wireless environment even though multiple channels can be used to reduce the transmission delay [29].

Furthermore, it was noticed in [17] that a page-based caching mechanism usually requires a high degree of locality among database items cached within a page in order for the mechanism to be effective. However, in [10], we show that the degree of locality in a mobile environment is usually very low. Since mobile clients are powered by short-life batteries [23]. Caching a page will result in wasting of energy when the degree of locality is low. Thus, the overhead of transmitting a page over a low bandwidth wireless channel for each query would be too expensive to be justified. Thus, it is necessary to consider caching at a smaller granularity in this mobile context.

2.1.2. Cache coherence. In a mobile computing environment, three cache invalidation mechanisms for cache coherence have been proposed [4], namely *Broadcasting Timestamps*, *Amnesic Terminals*, and *Signatures*. The connection frequency and duration of mobile clients (whether client is workaholic or sleeper) have been taken into consideration in these mechanisms. Extending the idea of using signatures for cache invalidation, Lee et al. propose three signature-based approaches, namely *simple signature*, *integrated signature*, and *multi-level signature* for efficient information filtering on broadcast channel [28]. Four signatures caching policies are also proposed, namely *Bit Tags and Active Refresh*, *Version Number and Active Refresh*, *Bit Tags and Passive Refresh*, and *Version Number and Passive Refresh*.

Conventional cache coherence strategies [7] require a server to be a stateful one. In other words, they require the server to have complete knowledge of each individual client's cache to be responsible for notifying all relevant clients whenever an item is updated. By contrast, the Leases file system [19] employs a different approach. Each file cached in the local storage of a client is associated with a pre-specified *refresh time* which defines the duration within which the cached file could be regarded as valid in the client's local storage. When the refresh time expires, the client needs to contact the server for an updated file.

2.1.3. Cache replacement. Whenever the cache storage is exhausted, a cache replacement algorithm should be employed to make room for the incoming database items by discarding previously cached database items [32]. In [16], various cache replacement algorithms including least recently used (LRU) and least reference density (LRD) have been proposed and their suitabilities in a conventional database system have been examined. LRU is subsequently generalized to LRU-*k* in [33], with an attempt to improve the "memoryless" property of LRU.

The *Coda File System* [26, 36] is designed and implemented employing *Prioritized Cache Management* to maintain mobile clients' *MiniCache*. The files to be cached are selected according to information of recent reference history (one kind of LRU) and user-specified information stored in *hoard database*, suitable in a mobile environment.

We believe that historical information of access frequency is important to identify the "hot spot" in a database. LRU may not be as appropriate in this way. Adaptation is also an important property of a replacement algorithm so that when the access pattern on database items changes, such as those location-dependent information with respect to a client, the cache replacement algorithm should be able to adapt to such changes. It is therefore necessary to re-address the issue of cache replacement in this context.

2.2. *Consistency of database*

The benefits of data caching must be balanced against the additional cost and complexities introduced for maintaining the consistency of the database. The consistency of the database is determined by the definition of *correctness criterion*, which is often the provision of transactions [5] to access the database. Concurrency control protocols are required to synchronize concurrent transactions accessing the database to ensure the defined correctness criterion.

The widely accepted correctness criterion for a set of concurrent transactions is *serializability* [5] in which there is a legal serialization order for the set of transactions, satisfying the ACID properties. There are numerous concurrency control protocols for transaction processing in conventional distributed environment. Basically, they can be classified into two approaches, namely *pessimistic* and *optimistic*, employing two major synchronization techniques, namely *lock-based* and *timestamp-based*, to construct a protocol [13].

In a mobile environment, the optimistic approach appears to form the mainstream of protocols for transaction processing [1, 20, 26, 36]. They require user-specified information such as the user-specified *Application Specific Resolver* (ASR) in Coda [26] and the high-level semantic information of objects in Thor [20]. On the other hand, lock-based approach is adopted in the Shore database with certain success [8].

Ensuring serializability is often too expensive in a database system where the communication delay is large and transactions are executed for a long period. Enforcing serializability in a mobile environment is impractical, due to similar reasons. There has been extensive research in allowing a controlled degree of inconsistency for read-only transactions to improve the concurrency and efficiency in transaction processing [27, 34, 38]. The usefulness of these alternative notions of correctness criterion together with appropriate concurrency control protocol should be investigated in the mobile environment.

3. The mobile caching model

We briefly describe our mobile caching mechanism in the framework of MODEC. Implementation of the prototype will be described in Section 4. We need to cache database items in the local storage of a mobile client and we choose to model that as a local OODB, using standard ODMG constructs [3]. We investigate three different levels of granularity for caching a database item, namely *attribute caching*, *object caching*, and *hybrid caching* [10].

Several design issues need to be addressed. First, a cache table is required in each client to identify if a database item is cached in local storage. Second, if a client is connected to a server, the client should be able to retrieve the cached items from local storage and the uncached items from the server. The client will only retrieve the cached items if the network is disconnected. The subtle differences in data accessing procedure should be transparent to a mobile client, except that the client should be notified with a return status distinguishing whether the server has been contacted. Third, an effective coherence strategy is needed to maintain the freshness of the cached items. Fourth, an effective replacement algorithm must be identified to retain the most frequently accessed items. Finally, a concurrency control protocol is needed to maintain the correctness of cached items.

3.1. The cache table

A cache table is maintained by a mobile client to indicate whether a database item is cached in its local storage or merely resides at the server. It is a mini-database, which could be managed by the OODB management system at the client. A class hierarchy rooted at class **Remote** and a class hierarchy rooted at class **Cache** are maintained. For each class **X** maintained at the server, a corresponding class **X** is created as subclass of **Remote** and a class **C_X** is created as subclass of **Cache** at a client. Class **Remote** indicates if an item is cached while class **Cache** holds the cached value of the item. For each attribute, a, of class **X** at the server, a user-defined method with the same name, a, will be created for class **X** in the client's local storage. This method encapsulates the tasks involved in the query processing model described in Section 3.2. In addition, an attribute with name c_a is created for class **C_X**, with the same definition as attribute a at the server.

To illustrate, consider an Advanced Traveler Information System (ATIS) application [12] in which a group of tourists would like to retrieve places of attraction and accommodation information. Figure 1(a) illustrates a simplified OODB for this traveler information system while figure 1(b) indicates the (partial) structure of the local database maintained by a mobile client.

To cache an object, x, belonging to class **X**, a local *surrogate* object is created for **X** in the client's local database. Each local surrogate will inherit two attributes from class **Remote**: R_oid holding the OID used by the server to reference x and R_host holding the server address where x originally resides. The local surrogate x is added to class **C_X** via multiple membership modeling construct of OODB model, inheriting attributes defined for **C_X** which are used as placeholders to cache attributes of x. The value of an attribute, a, of an object, x, $a(x)$, will be cached under c_a of **C_X**. For instance, in figure 1(b), a local

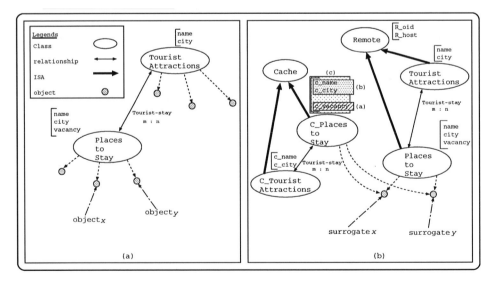

Figure 1.   A sample ATIS database application in mobile caching.

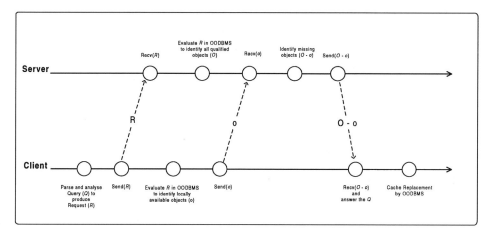

Figure 2. Query processing in the system prototype.

surrogate, referenced as x, is created as an instance of **Places to Stay**. Surrogate x is further added to **C_Places to Stay** via multiple membership construct [10]. The hierarchy rooted at **Remote** functions as a cache table, indicating those attributes of objects cached in the local storage.

3.2. Query processing

The sequence of execution events for query processing in MODEC is shown in figure 2. When a client initiates a query, Q, which tries to access attributes of a class, **X**, the user-defined method for the relevant attribute of **X** in the local hierarchy will send a request, R to the server. Both client and server proceed to evaluate Q concurrently. The client will attempt to evaluate Q by retrieving qualified attributes from the **Cache** hierarchy. An existent list containing the identities of the locally cached database items is resulted after evaluation and will be subsequently sent to the server. The existent list reports to the server about those database items which have been qualified and cached locally so that they do not need to be transmitted. The server will identify those missing database items and send them to the client. The client could, thus, respond to Q and cache the newly obtained items in its local storage. If the local storage is exhausted, the cache replacement algorithm will be invoked to identify the victim items to be discarded.

3.3. Cache granularity

The cache table provides a uniform framework to support caching at multiple granularity in MODEC, namely attribute caching, object caching, and hybrid caching.

In *attribute caching*, after the server has evaluated the query submitted by a client, it only returns those attributes of those qualified objects requested by the client. To illustrate, consider the following OQL [3] query, Q, being evaluated against the ATIS database in figure 1(a):

```
select x.name, x.city
from x in Places to Stay
where x.vacancy > 0.
```

With the client's local database schema as shown in figure 1(b), assume that before the query is initiated, the client only has surrogate x and attribute vacancy of x is cached (shaded region (a) in figure 1(b) and that only two objects, x and y, maintained at the database server satisfy Q. When evaluating the where clause of Q, the method vacancy of Places to Stay will submit the query to the server. Concurrently, it will also retrieve vacancy(x) from c_vacancy(x) and submit an existent list to the server indicating that vacancy(x) has been cached. The server will return a list containing values for name(x), city(x), name(y), city(y), and vacancy(y). The client can then cache name and city, of surrogate x under c_name and c_city respectively, as indicated by shaded region (b) in figure 1(b). A new surrogate for object y will also be created for caching the relevant attributes of y.

Each mobile client usually has its own set of hot objects. However, depending on the queries, it might access different attributes of an object requested earlier. When the server receives a request, it might be worthwhile for the server to send all attributes of a qualified object, x to eliminate future requests for x. This idea of *object caching* is to prefetch all attributes of a requested object to the initiated client, which will cache the returned attributes under class C_X as in attribute caching. For instance, the region (c) in figure 1(b) shows that all attributes of x will be cached when query Q is evaluated.

In object caching, the database server will send all attributes of a qualified object, x to the client. It is very often that not all attributes of x will be accessed in the future. This not only wastes the transmission bandwidth, but also occupies storage for caching other more frequently accessed attributes. *Hybrid caching* restricts the database to prefetch only those attributes of a qualified object with a high likelihood to be accessed in the future. Only those attributes whose access probabilities are above a *prefetching threshold*, ϵ, will be prefetched.

3.4. Cache replacement

In our design of MODEC, we have considered a number of replacement algorithms that make use of the access probabilities of database items. We illustrate the idea in the context of object caching; nonetheless, replacement algorithms under attribute caching and hybrid caching are similar in nature. For each database object, x, a *replacement score* indicating the prediction of its access probability is estimated. The replacement score is computed by a *scoring function*. Let us define a few notations to be used throughout this paper. The measure that contributes to the replacement score, used in the scoring function, for an object x at time t is denoted as $M_{x,t}$. The replacement score at time t for x is denoted as $\bar{M}_{x,t}$. Any superscript to the notation indicates the parameter in the appropriate algorithm. There are three different replacement algorithms with different scoring functions in MODEC, namely, Mean, Window and EWMA (Exponentially Weighted Moving Average). Let us briefly review them here.

The Mean scoring function measures either the average number of accesses per unit time or the average inter-operation arrival duration. The replacement score for each object x can be calculated incrementally by the scoring function

$$\bar{M}_{x,n+1} = \bar{M}_{x,n} + (M_{x,n+1} - \bar{M}_{x,n})/(n+1).$$

With a stable access pattern, Mean is expected to perform well. On the contrary, Mean responds slowly to changes in access pattern due to the limited effect of new accesses on $\bar{M}_{x,n+1}$ with large n.

The Window algorithm aims at improving the drawback of Mean. Borrowing the idea of *working set* [14], the mean inter-operation arrival rate within a window of period (W) is measured for each object as replacement score. The replacement score for x is estimated as:

$$\bar{M}_{x,n+1}^{(W_x)} = \bar{M}_{x,n}^{(W_x)} + \left(M_{x,n+1}^{(W_x)} - M_{x,n-W_x+1}^{(W_x)}\right)/W_x.$$

For practical implementation of Window, we need to reduce the amount of storage required in maintaining the W_x intermediate values. This is achieved by assuming that the accesses within the last window frame are evenly distributed. When calculating the new score, the term $M_{x,n}^{(W_x)}$ can be replaced by $\bar{M}_{x,n}^{(W_x)}$. Nevertheless, this assumption may result in less adaptive behavior on the algorithm since the weighting of the new score on the new mean value is averaged over the measures within the last window frame.

To adapt quickly to changes in access patterns, we assign exponentially decaying weights to different measures in the EWMA algorithm. One important parameter in EWMA is the exponentially decreasing weight, α, which ranges from 0 to 1. For object x with weight α_x, the estimated replacement score for x is computed as:

$$\bar{M}_{x,n+1}^{(\alpha_x)} = \alpha_x \bar{M}_{x,n}^{(\alpha_x)} + (1 - \alpha_x)M_{x,n+1}.$$

Notice that the value of α is application-dependent and can be obtained experimentally [9].

3.5. Cache coherence

Cache coherence involves cache invalidation and update mechanisms to invalidate and update out-dated cached items, respectively.

Since most applications in a mobile environment will generate more read operations than write operations [21], a mobile client could often accept a slight degree of out-dated data in return for faster data retrieval. In MODEC, we adopt a "lazy pull-based" invalidation approach in which each client is responsible to invalidate its own cached items and an on-demand update approach in which a stale cached item is only updated on its next access. We illustrate the idea using object caching; cache coherence in attribute caching and hybrid caching are similar.

Each cached object, x, is associated with a *refresh time*. The refresh time for x, RT_x, indicates the duration that x could be assumed to be valid in a client. In attribute and hybrid caching, since individual attributes of x are cached, the estimated refresh time is associated with individual attributes. When a client accesses x, it checks the validity of x by determining if RT_x has expired. When the server returns x in response to the request, it will estimate the refresh time for x, to be sent along with x to the client. This approach does not require a client to be always connected in order to invalidate x. Furthermore, if x is never accessed again, x will never be refreshed even though its refresh time expires.

We employ two methods for estimating the refresh time: *standard statistical estimation method* and *adaptive statistical estimation method*. The inter-arrival duration of consecutive write operations, d_x, for each cached object x is maintained. We estimate the refresh time for object x, RT_x, by $\bar{d}_x + \beta_x s_x$. The value of the adjustable parameter, β_x, determines the relative frequency of refreshing x. The smaller the value of β_x, the smaller the refresh time and the higher the possibility that the client needs to request x when it accesses x in a query. In the standard method, $\bar{d}_x$ and s_x are the mean and standard deviation of inter-arrival durations of write operations respectively.

The adaptive method shares the spirit and lies along the same vein as EWMA. We extend the estimation of the effective mean and effective standard deviation by assigning exponentially decreasing weights to each statistical quantity of the update duration and measure so that $\bar{d}_x$ and s_x are the exponentially weighted measure of mean and standard deviation. With parameter γ_x, the exponentially weighted mean, i.e., EWMA, will be $\bar{d}_{x,n+1} = \gamma_x \bar{d}_{x,n} + (1 - \gamma_x) d_{x,n+1}$. The variance of the durations, which we call *Exponentially Weighted Moving Variance* (EWMV), is

$$Var^*_{x,n+1} = \frac{\sum_{i=1}^{n+1} \gamma_x^{2(n+1-i)} (d_{x,i} - \bar{d}_{x,n+1})^2}{\sum_{i=1}^{n+1} \gamma_x^{2(n+1-i)}}.$$

This can be computed incrementally as $Var^*_{x,n+1} = \bar{S}_{x,n+1} + \bar{d}^2_{x,n+1} - 2\bar{t}_{x,n+1}\bar{d}_{x,n+1}$, where $\bar{S}_{x,n+1}$, and $\bar{t}_{x,n+1}$ are

$$\bar{S}_{x,n+1} = \gamma_x^2 \bar{S}_{x,n} + \left(1 - \gamma_x^2\right) d^2_{x,n+1},$$

$$\bar{t}_{x,n+1} = \gamma_x^2 \bar{t}_{x,n} + \left(1 - \gamma_x^2\right) d_{x,n+1},$$

with initial values $\bar{d}_{x,1} = \bar{t}_{x,1} = d_{x,1}$, $\bar{S}_{x,1} = d^2_{x,1}$. The standard deviation s_x is the square root of the variance.

To measure the degree of coherence achieved, we define the notion of *error* and *false alarm* in accessing an object. Assume that a mobile client refreshes an object, x, at time t_1 and t_2. Between t_1 and t_2, the client might issue read operations on its local cached copy of x. For each read operation, $R(x)$, initiated between t_1 and t_2, if the server performs a write operation, $W(x)$, before $R(x)$ in real time, the value of x used by the client would be inconsistent with the actual value maintained at the server and hence, $R(x)$ results in an error. False alarm measures useless refreshes that consume bandwidth unnecessarily. If a client refreshes x at time t_1 and t_2 but there exists no write operation on x at the server during the period, the refresh at t_2 is useless. The definitions of error rate and false alarm are used in characterizing the performance versus coherence tradeoff in Section 5.

3.6. Transaction processing

Most applications in mobile environment will generate more read operations than write operations [21]. In other words, a transaction issued by a mobile client is likely a read-only transaction. It is often accepted that read-only transactions may tolerate a certain degree of inconsistency. Thus, the strictness of serializability may impose unnecessary limitations to system throughput [34]. We propose to use ϵ-serializability (ESR) as the correctness criterion for transaction processing in our MODEC environment. Integrating with the idea of *bounded ignorance*, our transaction model is capable of handling textual data as well. We also notice that pessimistic approach may not be a suitable approach for concurrency control in a mobile environment since a wireless communication link is subject to frequent disconnection.

3.6.1. Concurrency control protocol.

We employ optimistic approach to construct the concurrency control protocol in MODEC, similar in spirit to Coda [26] and Thor [20]. The principal reason is that the number of messages required to synchronize transactions in optimistic schemes is sufficiently smaller than that in pessimistic schemes. Moreover, the low contention environment assumption of optimistic approach is highly realistic in a mobile environment. We adopt backward validation to synchronize concurrent transactions when they commit. Each transaction, T, is divided into four execution phases: *Read*, *Computation*, *Validation*, and *Write*. The protocol is described as follows.

1. In *Read* phase, data requests from a mobile client are satisfied by the local cache and remote server. There is no lock required prior to access of these data. Instead, the transaction manager (TM), which is installed at the client, will log all data accesses in a transaction log ($T_{\log}$).
2. During *Computation* phase, the data are manipulated and possibly updated in mobile client local storage only. There is no action taken by the server. All activities are logged in $T_{\log}$ by the TM.
3. When T attempts to commit, TM will forward $T_{\log}$ to the server. Upon reception of $T_{\log}$, the server will attempt to validate $T_{\log}$ in the *Validation* phase. The transaction T corresponding to $T_{\log}$ will be committed if and only if $\forall x \in T_{\log}$, $x_{\log}.WC$ and $x_{base}.WC$ are identical in value, where WC is the *write counter*, which keeps track of the number of write operations made on the database item x, while $x_{\log}$ and x_{base} are the copy of x in the log and in the server database respectively.
4. If a transaction T is ready to commit, all writes in $T_{\log}$ are installed at the server database during the *Write* phase. Meanwhile, each write operation on x_{base} will cause $x_{base}.WC$ to be increased by one. When a mobile client receives the commit message from the server, the corresponding transaction will be committed locally. The $T_{\log}$ kept by the TM will be discarded accordingly. Otherwise, T will be aborted.

5. Whenever T is aborted, the corresponding $T_{\log}$ kept by the TM and all modified data by T in client's cache will be discarded accordingly.

It is noted that during validation phase, the server performs validation within a critical section to obtain a consistent snapshot at the server database. To allow concurrent updates at the server, we employ two phase locking protocol in modifying the database items updated by a transaction. Since all data accesses are known in prior, it is possible to acquire locks in an incremental manner following the identifier of the database items without causing deadlock. All concurrent transactions, therefore, are serialized at the server during validation. In other words, serializability of update transactions is enforced in this phase.

3.6.2. Bounded ignorance and ESR.

Referring to the third step of the protocol in Section 3.6.1, the condition $\forall x \in T_{\log}$ such that $x_{\log}.WC = x_{\text{base}}.WC$ can be modified to allow certain inconsistency so as to improve the system throughput. In fact, it can be re-written as $\forall x \in T_{\log}$ such that $|x_{\log}.WC - x_{\text{base}}.WC| \leq nLimit$, where $nLimit$ is a pre-defined inconsistency allowance value. This condition specifies that a transaction may ignore the $nLimit$ version(s) of x written by previous $nLimit$ committed transaction(s).

For numeric data, a more precise inconsistency in term of numeric value can be defined by users. We adopt the ESR model in our transaction model. In case where x is numerical, the term $x_{\log}.WC$ has to be replaced by $x_{\log}.nWC$. For each attribute of a class, a pre-defined $ImpLimit$ is associated. Thus, the value of $x_{\log}.nWC$ would be

$$x_{\log}.nWC = \begin{cases} |x_{\log} - x_{\text{base}}|/ImpLimit_x & \text{if } ImpLimit_x > 0 \\ 0 & \texttt{otherwise.} \end{cases}$$

Our protocol is similar to the *Divergence Control Protocoli* for ESR [34]. The inconsistency only applies to read-only transactions but not to update transactions, which must be executed in a serializable manner in order to preserve the consistency of the database.

4. System architecture

In order to evaluate the performance of MODEC, we have developed a simulation model followed by a system prototype implementation. The simulation model and a comprehensive list of simulated experiments conducted have been discussed in [10]. We focus our discussion on our prototype in this paper. The system prototype is developed following a simple client-server architecture as depicted in figure 3.

In the prototype, both client and server are installed with an ODE *Object-oriented Database Management System* (OODBMS). A *Communication Module*, which enables communication between the client and server, is integrated with each OODBMS. The server-OODBMS is responsible for the maintenance of the server database and handling object requests from clients. The client-OODBMS is responsible for the maintenance of the local cache, requesting the server in case of cache missed, and answering queries from applications. In order to support OQL suggested by ODMG [3], a *Query Parser* (QP) is implemented upon the OODBMS at the client. The QP is responsible for parsing and analyzing OQL queries from applications and converting them into a low-level language

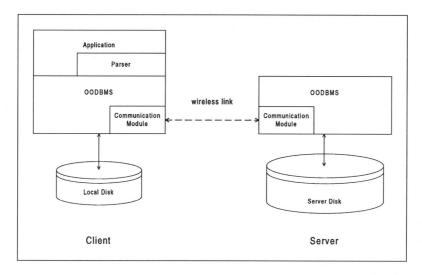

Figure 3.   System prototype architecture.

understandable by ODE. The QP is implemented as a separate module of the OODBMS. This allows the QP to be portable among different OODB engines.

4.1. Data modeling in sample application

We use a real-life database in our experiments. The database is obtained from the Land Department, Government of the Hong Kong Special Administrative Region. The database maintain information on real-estates in Hong Kong. All data in this database are stored in relational database format. It is necessary to transform the database schema to OODB model for our implementation. The revised object-oriented schema definition in form of a class hierarchy is depicted in figure 4.

In our sample application, there are only two object classes, Flat and Xtran, both being subclasses of the root class Cache. This root class implements the client caching mechanism, including replacement score and access time associated with each database item. Flat and Xtran exhibit a simple one-to-one relationship: attribute xtran of class Flat is a reference to an object of class Xtran, whose attribute flat is also a reference to an object of class Flat. In our database, we create 2000 data objects containing real-life data.

We implement an object-query language (OQL) interpreter (i.e., the QP) following the ODMG [3] standard using Lex & Yacc utilities (PC version) [31]. Instead of implementing all OQL constructs suggested by the ODMG, we only implement a subset of it, enough for the purpose of verifying our concepts in this context. An example of an OQL query on our sample database will be

```
select f.BldName, f.Size, f.xtran->UPrice
from f in Flat
where f.xtran->InsDate = "1997/06/27".
```

```
//
//     Cache class definition
persistent class Cache {
  public:
  indexable
    unsigned int    OId;             // object ID
    float           S;               // replacement score
    double          LastUpdateTime;  // last update time
    double          ExpTime;         // expiry time stored in client
    unsigned int    WTS;             // write timestamp
};
//
//     General information of house and apartment
persistent class Flat : public Cache {
  private:
    char         Unit[10];          // unit number
    char         Floor[10];         // floor number
    char         BldcName[50];      // building name (Chinese)
    char         BldName[80];       // building name (English)
    char         BlkNo[10];         // block number
    char         EstcName[50];      // estate name (Chinese)
    char         EstName[80];       // estate name (English)
    char         StNo1[8];          // beginning street number
    char         StNo2[8];          // ending street number
    char         StcName[50];       // street name (Chinese)
    char         StName[80];        // street name (English)
    char         Dcode[12];         // district code (by government)
    char         SizeG[10];         // size (gross)
    char         Size[10];          // size (sellable area)
  persistent
    Xtran        *xtran;            // history of buy/sell transactions
};
//
//     History of property buying and selling
persistent class Xtran : public Cache {
  private:
  persistent
    Flat         *flat;             // general information of the property
    char         MemNo[20];         // memorial number
    char         InsDate[12];       // date of signing contract (format: "yyyy/mm/dd")
    char         ComDate[12];       // date of completion
    char         DelDate[12];       // date of delivery
    char         PropKind[10];      // (e.g. RES, IND...)
    char         Nature[80];        // (format: P-ASP, ASP, ASSIGN, P/AUCT + desc)
    char         ConPrice[10];      // consideration price
    char         UPriceG[10];       // unit price gross
    char         UPrice[10];        // unit price
    char         Assignee[80];      // who buy  (company or individual or etc)
    char         SoldBy[80];        // who sold (company or individual or etc)
};
```

Figure 4. Class definitions for sample application.

4.2. *Pragmatic issues in prototype development*

Since the physical addresses of objects in the server and client are different, accessing the reference pointer of an object in a cache involves *pointer swizzling* [25], which is implemented with a copy, lazy and direct approach in our prototype. When an object is copied from the server, the pointer in the referring object will be updated to the physical address of the referenced object (direct reference), only when the reference pointer in an object is accessed on-demand. The pointer swizzling mechanism is encapsulated in the QP.

It is noted that the *maximum transmission unit* (MTU) in Ethernet is limited to 1500 bytes [37], excluding the 14-byte Ethernet header and the 4-byte Ethernet trailer. Conceptually, with the TCP/IP protocol, the IP layer should handle message fragmentation and reassembly. In practice, the assumption is valid only when both server and client are running on the Unix platform. Since our client program is running on the PC platform (see Section 5), the server sending a message exceeding an MTU may be fragmented into several messages. We discover that fragmentation occurs when the transmission unit exceeds 1460 bytes. We thus have to synchronize the send and receive operations with sequence number control for large messages. The application communication protocol includes the length of data stream in the first 4-byte of a message to be sent, so that reassembly of fragmented messages can be performed at the communication module of the receiver.

5. System performance

We have conducted extensive simulation studies [10] on mobile data access based on the MODEC mechanism. In this paper, we select some representative simulation results as well as enrich our simulation studies to include the allowance of limited inconsistency in mobile transactions for better performance. In addition, we conduct the same experiments on the prototype to validate our simulation results. For our experiments on the prototype, we try to maintain the same experimental settings as those we used in our simulation so that our simulation results can be used to cross-validate with those from prototype. Each experiment on the prototype is run for two to four days.

In each experiment, one server and 10 clients are involved. Two wireless communication channels (one for uplink and one for downlink), are shared among the 10 clients. We are using Digital Roamabout wireless equipment [15]. The bandwidth of a wireless channel is about 19.2 Kbps. The database server runs on a Sun UltraSparc workstation while the client database runs on a 133 MHz Pentium PC. The server's OODBMS maintains 2000 database objects, while each client is able to maintain 400 database objects in its local storage. The memory caches of the server and each client are maintained by their respective operating systems. We experiment with a *changing skewed heat* (CSH) access pattern. We adopt an 80/20 rule in defining the hot objects: 20% of the objects are picked randomly such that 80% of accesses to objects fall within this subset. The hot objects are reselected after every 500 queries.

Two types of query, *associative query* and *navigational query*, are considered in our experiments. In *associative query* (AQ), Q_a primitive-valued attributes of each selected object will be accessed. The attributes of each object that will be selected by a query follow a uniform skewed distribution. All attributes have a non-zero access probability. In *navigational query* (NQ), for each selected object, x, $\mathcal{Q}_n$ inter-object relationship of x will be traversed in additional to accessing $\mathcal{Q}_a$ primitive-valued attributes of x. When traversing the relationship, $\mathcal{Q}_a$ attributes of the related object, will also be referenced. The access operations arrive in a Poisson pattern.

5.1. *Performance evaluation*

Our experiments are organized around four objectives. First, we would like to study the performance differences with respect to cache granularity, comparing attribute caching (AC), object caching (OC), and hybrid caching (HC) with the case of no caching (NC). Second, we would like to compare our replacement algorithms with the conventional ones. Third, we would like to study the effectiveness of our coherence strategy in maintaining the freshness of cached database items. Finally, we would like to study the impact of concurrency control protocol on the performance of our caching model.

Within the graphs that we will be presenting in the coming subsections, we have arranged the performance results from simulation and from prototype side by side for comparison. This cross-validation step is rather successful and we believe that the simulation results can show the *relative* performance of our various approaches under a variety of parameter settings rather accurately. Besides showing that the MODEC mechanism is useful in the mobile prototype, the more cost-effective simulation studies can be adopted to study new protocols and algorithms designed under a similar operating environment, before being implemented and evaluated empirically.

There are five measurements observed for performance evaluation including *Response Time, Hit Ratio, Error Rate, False Alarm*, and *Commit Rate*. Since there is no information about the overhead for OQL evaluation, the response time measured in simulation model does not include the overhead for OQL evaluation. Thus, for the experiments running on the system prototype, *Response Time* ($\mathcal{RT}$) is defined as: $\mathcal{RT} = t_2 - t_1$, where t_1 is the time recorded by the QP just before sending the request to the server, and t_2 is the time recorded by the application after receiving the corresponding response. In addition, the OQL evaluation time ($\mathcal{ET}$) which is defined as: $\mathcal{ET} = t_3 - t_2$, where t_3 is the time recorded by the application just before supplying query to the QP.

Hit ratio ($\mathcal{HIT}$) is defined as: $\mathcal{HIT} = \#ObjectAccess_{cache}/\#ObjectAccess$ where $\#ObjectAccess_{cache}$ is the total number of accesses satisfied by non-expired items in local cache, while $\#ObjectAccess$ is simply the total number of accesses on items. It is noted that our definition of $\mathcal{HIT}$ is different from the one employed in conventional memory caching [18]. In conventional memory caching, there are four states for each cached item:

Invalid   Data is not available in the cache;

Valid   Data is available in the cache and not modified;

Reserved   Data is available in the cache and locally modified exactly once since it was brought into the cache and the change has been transmitted to backing store;

Dirty   The data has been locally modified more than once since it was brought into the cache and the latest changes has not been transmitted to backing store.

Only the **Valid** case is considered as a cache hit in conventional caching. In a mobile environment, mobile clients are ignorant of updates made in server especially when being disconnected. Therefore, we would like to re-define $\mathcal{HIT}$ to include dirty cache hit as an indicator for the data availability of a mobile client. In addition, we measure the error rate ($\mathcal{ERR}$) to capture the effects of dirty cache as defined in [18], defined as: $\mathcal{ERR} = \#ObjectAccess_{dirty\_cache}/\#ObjectAccess_{cache}$.

False alarm ($\mathcal{FA}$) is defined as: $\mathcal{FA} = \#Refresh_{useless}/\#Refresh$, where $\#Refresh_{useless}$ is the total number of useless refreshes (see Section 3.5), while $\#Refresh$ is the total number of refreshes made on an expired cached item. It indicates the effectiveness of the estimated refresh times for refreshing cached items and consequently reflects the utilization of wireless bandwidth for maintaining the refreshness of cached items during the period of connecting to the server.

Finally, we measure the commit rate of the system to characterize the throughput of the system regarding its transaction processing ability. Commit Rate ($\mathcal{CR}$) is defined as: $\mathcal{CR} = \#Commit/(\#Commit + \#Abort)$, where $\#Commit$ is the number of committed transactions, and $\#Abort$ is the number of aborted transactions.

5.2. Experiment A

The first set of experiments on the prototype aims at studying the performance of AC, OC, HC, and NC for both AQ and NQ. We use $\mathcal{Q}_a = 3$ in AQ and $\mathcal{Q}_n = 1$, $\mathcal{Q}_a = 1$ in NQ. As a result, the selectivity and also the size of the result sets in both AQ and NQ are the same. However, the number of objects assessed in NQ would be twice that number assessed in AQ. The prefetching threshold for HC is set to two standard deviations below the mean access probabilities of all attributes. NC is achieved by disabling storage cache on all clients. For storage caching, EWMA with α equal to 0.5 (EWMA-0.5) is employed as the replacement algorithm, since we have demonstrated that $\alpha = 0.5$ is a judicious choice in EWMA [9]. The update probability is fixed at 0.1. We use $\beta = 0$ in both statistical refresh time estimations; therefore, statistical estimation methods would behave very similarly, since our EWMV method is primarily targeted on the variance. The hit ratios, response times, error rates, and false alarms are measured. We do not measure the effect of concurrency control in this experiment. It will be discussed in more detail in Experiments D and E.

The results are depicted in figure 5. The first set of graphs (figures 5(a)–(d)) illustrates the performance of AQ while the second set of graphs (figures 5(e)–(h)) illustrates the performance of NQ. Within each graph, we present the performance obtained through simulation side by side with that obtained from experiments on the prototype for comparison purpose.

Note that the hit ratio, error rate, and false alarm for NC are not available in prototype experiments because we are not able to measure them accurately since memory caching is handled by the operating system. The characterizing performance measurement for the NC is therefore the response time.

From figure 5, it is clear that the base case (NC) performs a lot worse than any storage caching scheme. It has a much lower hit ratio (from simulation) and a much higher response time than those of any storage caching scheme. This is mainly because a storage caching scheme trades network transmission for local disk access which has a much higher bandwidth than that of a wireless channel.

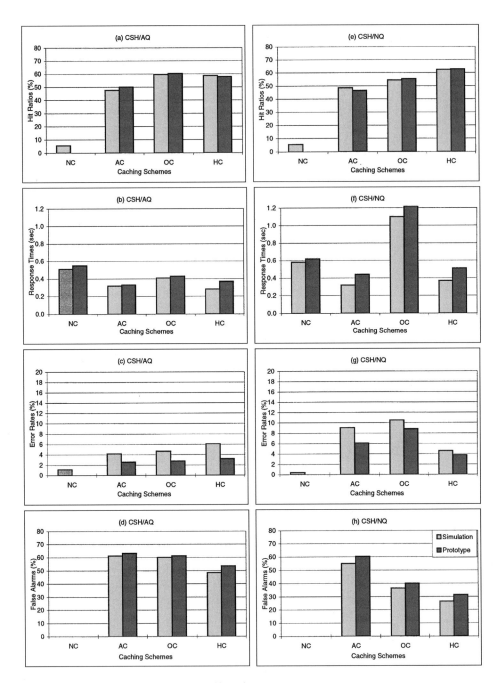

Figure 5. Performance of various storage caching schemes.

We observe that OC, both in simulation and prototype, yields higher hit ratios than AC. This is mainly because OC will cache all attributes of hot objects, thus eliminating extra requests when different attributes of the same object are requested in the future. It is interesting that despite the higher cache hit ratios from OC when compared with those from AC, OC results in higher response times as well (see figures 5(b) and (f)). This anomaly is mainly due to "blind prefetching" in OC. Since OC prefetches all attributes of an object, the overhead for transmitting cold attributes of hot objects will be wasted. The overhead saved due to the high cache hit is offset by the even higher transmission delay due to the low bandwidth of wireless channel. HC performs very well, achieving response times as low as those of AC while achieving cache hit ratios close to those of OC. Prefetching in HC appears to be "intelligent", in that most attributes prefetched will, most likely, be accessed in the future.

When comparing results from simulation and prototype, we observe that the average hit ratios of various caching schemes achieved by the prototype are very close to those revealed by the simulation studies (within 2% difference). It supports the conclusion we make based on our simulation results [10] that data availability can be improved by data caching and further enhanced by prefetching mechanisms as in the case of OC and HC.

From figures 5(b) and (f), generally, higher response times for all caching schemes are observed in the prototype when comparing with those from simulation. The reason is that the simulation model measures the database access time based on the bandwidth of hard disks only. In practice, however, the access time depends also on the overhead due to database management system such as index table updates, garbage collection, etc. This accounts for the slightly higher response time.

NQ results in higher response times when compared with those of AQ. There are two main reasons for this observation. First, the number of objects accessed is double in NQ, even though we have enforced the same selectivity on the queries. Second, NQ requires more database management system overhead than AQ such as pointer swizzling and object traversal. As a result, response times in NQ generally are higher than those in AQ. We also observe the exceptionally high response times in case of OC. It is mainly due to "blind prefetching" of all accessed objects in NQ, at twice the amount as AQ.

In theory, the higher the refresh rate, the lower the error rate and possibly the higher the false alarm rate should be. This behavior can be observed in figure 5. We notice that error rates are slightly lower in the prototype when compared with those in simulation. In the simulation, when a query arrives, the required objects available in the cache will be collected. There might be some "vulnerable" objects which are due to expire very soon. In the prototype, the collection process will have a minor delay after the query arrival due to the overhead required for query parsing and evaluation. During this period, the "vulnerable" objects might have expired already and may not cause errors.

5.3. Experiment B

In the second set of experiments, we study the performance of various replacement algorithms for storage caching including LRU, LRU-k, LRD, and EWMA. The prefetching threshold for HC is again set to two standard deviations below the mean access probabilities of all attributes. The parameter α for EWMA is set to 0.5 (EWMA-0.5) while for LRU-k, k is set to 3 (LRU-3 is shown to perform well in [33]). For LRD, the reference count of each database item is divided by 2 every 1000 seconds. Choosing 0.5 for α in EWMA has the closest intuitive meaning as dividing the reference count by 2. The update probability is fixed at 0.1. The number of servers and clients remain at 1 and 10, respectively. Hit ratios, error rates, response times, and false alarms are measured.

The results are depicted in figure 6. The first set of graphs (figures 6(a)–(d)) illustrates the performance of AQ while the other set (figures 6(e)–(h)) illustrates the performance of NQ. Within each graph, the performance obtained from simulation is presented in parallel with that obtained from the prototype for comparison.

As observed in figure 6(a), the hit ratios achieved by applying different replacement algorithms are very close to the results obtained through simulation. The deviations between results from simulation and prototype experiments are at most 4%. The results also show that the *relative* hit ratios achieved by various cache replacement algorithms respect those obtained in simulation. LRU and LRU-3 perform very closely while LRD performs slightly better than LRU and LRU-3. EWMA-0.5, in turn, performs slightly better than LRD. Regarding the response times depicted in figure 6(b), a general rise is observed which is consistent to the observation in Experiment A. Error Rates and false alarms also exhibit the same behaviors as those observed in Experiment A.

The practical value of storage caching is proved by the prototype to be promising. Employing EWMA-0.5 as the replacement algorithm can achieve a reasonably higher performance in terms of hit ratio and response time when compared with conventional replacement algorithms.

5.4. Experiment C

In the third set of experiments, we investigate the effect of update probability on the performance of caching schemes. We experiment AC, OC, and HC with update probabilities ranging from 0.1, 0.3, to 0.5. We do not present the results for higher values of update probability since we believe that, in a mobile environment, the number of write operations should be less than the number of read operations. The hit ratios, error rates, response times, and false alarms are measured.

The results are depicted in figure 7. The first set of graphs (figures 7(a)–(d)) presents the simulation results while the other set (figures 7(e)–(h)) presents the results from the prototype. In general, OC and HC result in higher hit ratios when compared with AC, consistent to the observation in Experiment A.

In general, the hit ratios decrease as the update probability increases while the response times increase accordingly. This is because when the update probability increases, there will be more write operations on each database item. This will reduce the refresh time of a database item, thus, shortening the duration that an item could be used at a client's local storage. The increase in response times is mainly due to the decrease in hit ratios. When comparing the results from simulation and prototype, we can observe only at most 5% deviation in hit ratios.

We also observe that OC leads to higher error rates than those of AC and HC. This could be explained as follows. Assume that a client reads attribute a of a cached object, x, at

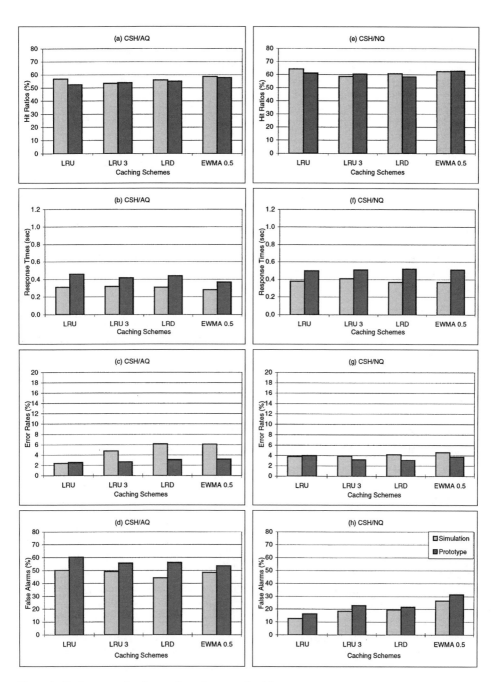

Figure 6. Performance of various cache replacement algorithms.

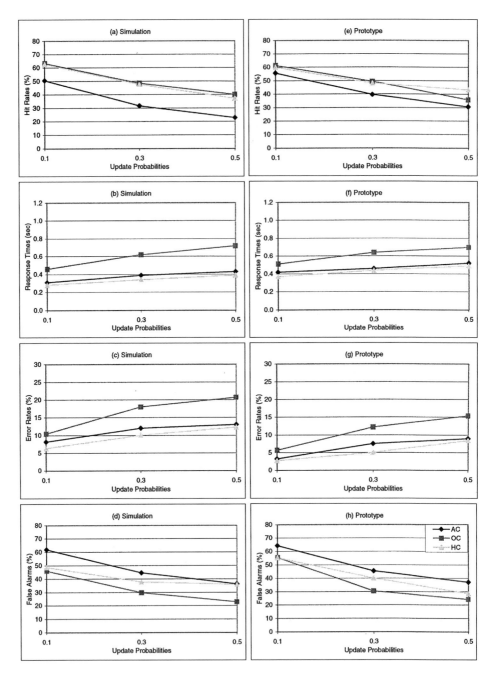

Figure 7.   Impacts of various update probabilities on system performance.

time t_1 and t_2 while attribute b of the same object is updated by another client at time t ($t_1 < t < t_2$), at the server. For OC, since the update of attribute b is considered as an update on object x which is the same object as the one being read at time t_2, the read operation at time t_2 is considered as an error. For AC and HC, the read operation at time t_2 will not be considered as an error since attributes a and b are considered as different items.

AC and HC result in similar number of errors. However, HC is able to satisfy more read operations using locally cached items due to the higher hit ratios. Hence, the error rates of HC are slightly lower than those of AC.

It is observed that AC results in the most number of false alarms while OC results in the least number of false alarms. HC, in general, results in a moderate number of false alarms. The reduction in false alarms by OC and HC is mainly due to their prefetching capability. Note that both OC and HC return extra attributes in addition to those explicitly requested by a client. For each prefetched attribute, if it has already been previously cached in the client's local storage, its newly prefetched value and its newly estimated refresh time will replace its old value and refresh time. This has an effect of extending the lease.

To study the improvement brought about by the adaptive method for refreshing stale database items, we conducted simulated experiments using $\beta = 1$. With a changing access operation arrival rate, we are able to observe the usefulness brought by the adaptive method. We briefly describe our observations here. With operations arriving under a Poisson pattern, both estimation methods show little difference. We subsequently model our update operations arrival pattern according to the Traffic pattern described in [24]. In brief, the Traffic pattern changes the arrival rate of update operations periodically. We first look at a simple situation where the set of hot objects remains unchanged across queries. We term such access pattern, *skewed heat* (SH). Under this Traffic pattern, the adaptive method demonstrates a distinctive advantage. With $\gamma = 0.5$, we discover that the hit ratios go up by 5 to 15%, from an average of 50% to an average of 60%. Meanwhile false alarms go from a range of 15 to 40% down to a range of 5 to 20%. As a result, the response times are reduced. The error rates, on the other hand, are only marginally increased by 1 to 5%. Repeating the experiment under the CSH access pattern, the reductions to false alarms and response times are even more significant, while the increases to hit ratios and error rates remain similar as in the case of the SH access pattern.

5.5. Experiment D

In this experiment, we study the performance of AC, OC, and HC with optimistic concurrency control (CC) protocol enabled to synchronize concurrent transactions issued among mobile clients. Each transaction is composed of an arbitrary number of queries, ranging from 3 to 8 queries. The results are depicted in figure 8. The first set of graphs (figures 8(a)–(d)) presents the results obtained without concurrency control while the second set (figures 8(e)–(h)) presents the results obtained with CC enabled.

It is observed that the resultant hit ratios with CC enabled are higher than those without concurrency control. It is believed that the increase in hit ratios is due to a longer estimated refresh time for each item. The reason for such an observation can be explained as follows. We assume that each transaction is composed of 3 to 8 queries. Updates to an item will

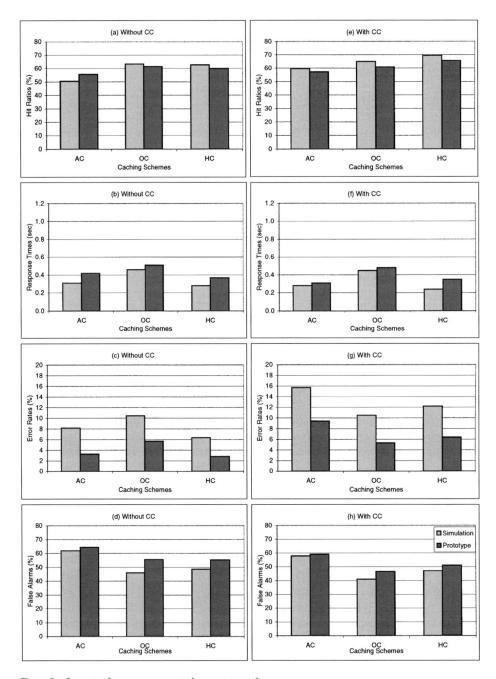

Figure 8. Impacts of concurrency control on system performance.

only have effect after a transaction is executed if the transaction is able to commit. If a transaction cannot pass the validation of the CC protocol, it will eventually be aborted and any update to objects issued by the transaction will be discarded. As a result, the average inter-arrival duration of update access for each item is longer when CC is enabled. Since the refresh time for each item is estimated according to its update rate, a longer inter-arrival duration will result in a longer estimated refresh time. This results in higher hit ratios and thus, lower response times.

On the contrary, we observe that the error rates are much higher when CC is enabled when compared with the case without CC. This observation is due to the longer refresh times estimated for items when CC is enabled. Therefore, there is a higher probability that an update operation will be committed between two refreshes. When comparing the results from the prototype and simulation, a general decrease in error rates is observed with increase in false alarms. This is consistent with the observation in Experiment A.

5.6. Experiment E

In our final set of experiments presented, we emphasize on the study of the system performance regarding the transaction processing ability. We repeat Experiment D with inconsistency tolerance (*Ignorance*) set to 0 and 1. The commit rates ($\mathcal{CR}$) are measured and the results are depicted in figure 9. The first set of graphs (figures 9(a) and 9(b)) presents the

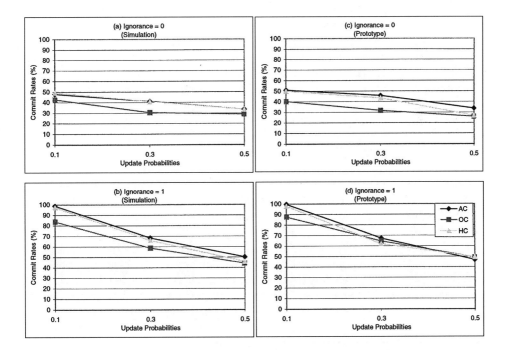

Figure 9. Performance of bounded-ignorance.

simulation results while the second set (figures 9(c) and (d)) depicts the results from the prototype.

It is observed that AC and HC result in similar commit rates which are higher than OC. It is due to a similar reason explained in Experiment C. For OC, updates to an attribute are regarded as updates to an object. Thus, the probability of conflicts among concurrent transactions issued by different clients is higher in OC when compared with the case in AC and HC.

Sustaining an inconsistency tolerance (*Ignorance* = 1) can significantly enhance the performance of transaction processing in terms of commit rates (see figures 9(b) and (d)). However, such improvement is restricted to low update probability. When the update probability is high, *Ignorance* does not have much effect on the commit rates. It is because inconsistency tolerance intends to relax only the consistency constraint of read-only transactions. The higher the update probability, the smaller number of read-only transactions can satisfy the restricted inconsistency constraint. Moreover, as update probability increases, the number of read-only transactions decreases significantly as well. Thus, inconsistency ignorance has less effect on commit rates. It is therefore, not beneficial to practice inconsistency tolerance in an update intensive transaction processing environment.

5.7. Discussion

We have conducted simulation studies on our mobile caching mechanisms and the results are verified empirically using our MODEC prototype, implemented with the ODE OODB with Unix database server and mobile PC client. We advocate the use of hybrid caching whenever possible, since it is yielding better performance in most cases than attribute caching or object caching alone, in term of the four performance metrics measured (revealed in Experiment A), whether concurrency control is enabled or not (revealed in Experiment D). It is also yielding the best performance regardless of the update probability in terms of hit ratio, response time and error rate, as demonstrated by Experiment C. In an attempt to improve performance in the presence of concurrency control by tolerating for controlled degree of inconsistency, hybrid caching also excels in allowing a higher transaction commit rate and hence system throughput, as evidenced in Experiment E. The more complete sets of simulation study results [9, 10] are in support of these observations. Intuitively, attribute caching only cache hot attributes to be used without prefetching for relevant attributes of the same object, leading to lower hit ratio. Object caching prefetches all attributes of any (hot) object selected for caching, leading to higher response time. Hybrid caching takes together the benefit of both in prefetching only hot attributes of hot objects, striking a balance successfully as we wished.

To further strengthen the benefit of hybrid caching, we observe that performance measured on our MODEC prototype verifies that the simulation results are applicable. The discrepancies between simulation and experimental results are either small, or deviated in the same direction with about same magnitude for different schemes in our experiments. The tradeoff of adopting hybrid caching scheme is perhaps a more complicated cache management scheme and the overhead in maintaining the replacement scores with respect to the more straightforward approach of object caching.

We also advocate the use of EWMA as an alternative to LRU for cache replacement unless the data access pattern is known, since EWMA is producing a rather good performance in terms of hit ratio and response time, the two more important performance metrics. LRU is also producing reasonable performance compared with the other replacement algorithms, as depicted in Experiment B empirically. As with LRU, EWMA is easy to use and easy to manage. Both LRD and LRU-3 have higher management overhead and their performances are not particularly good. Detailed studies on the replacement algorithms have shown that EWMA (and often LRU) have quite stable and relatively good performance across many different access patterns and that LRU-k and Window are able to produce good performance on several special data access patterns [9, 10].

We believe that the replacement algorithms are useful in the context of relational databases. Studies on broadcasting database items at the granularity of tuples over wireless channels showed that EWMA can lead to improved performance in terms of hit ratio over traditional LRU and MRU replacement algorithms [30]. Although hybrid caching can be generalized to relational databases by caching hot attributes of hot tuples of relations, it is not clear whether it can bring similar benefits, namely, higher hit ratio and lower response time, when compared with tuple caching (the equivalent of object caching) and attribute caching. This is perhaps due to the difference in query formulation. The join operation (the equivalent of navigational queries) in relational database requires a kind of an index on the joining attribute for efficient processing. To determine the set of tuples to be joined is not simple whereas in navigational queries, the related objects are traversed following the object identifiers in a unique way.

While LRU is useful in the context of file caching at a mobile client to cater for disconnected operation, EWMA would probably be useful in this context. Adaptive refreshing time estimation should serve as a more accurate metric for the invalidation or callback mechanism required by distributed file systems such as NFS, AFS [22] or Coda [26, 36]. This idea of adaptive refreshing has been applied in the context of web database pages, yielding a certain degree of improvement [35]. The use of EWMA in determining the virtual objects to be transmitted in a distributed virtual environment is also shown to be beneficial [11].

6. Conclusion

In this paper, we present a number of problems regarding data management and database access in a mobile computing environment. To address the limitations of narrow bandwidth and unreliable wireless communication links, we propose the MODEC mechanism for mobile caching as one way to improve performance of database access. We suggest three different levels of cache granularities, namely attribute caching, object caching, and hybrid caching in the context of object-oriented database. Based on access frequency of database items, we use several replacement algorithms, namely Mean, Window, and EWMA, each bearing different degree of adaptive behavior to changes in access frequency. We adopt a client pull-based strategy to maintain the freshness of cached items, through the notion of refresh time and two different statistical schemes for estimating the refresh time of a database item. We study the impact of transactional database access in a mobile environment and allow a controlled degree of inconsistency involved in a transaction by employing generalized bounded ignorance as the correctness notion.

Through a detailed simulation model, we found that the MODEC mechanism does improve the data access performance of mobile clients, especially with hybrid caching utilizing the **EWMA** replacement algorithm. Additional benefits can be achieved by using adaptive statistical estimation for refreshing stale cached items in the presence of varying query arrival rates. To validate the simulation results, we have developed a system prototype modeled after the ODMG constructs. The empirical results from our prototype are in strong supports of our findings in simulation, especially the relative performance among various algorithms. To evaluate the effectiveness of a new replacement algorithm, the algorithm can be studied under the simulation model in details and the result would likely be indicative in the practical environment.

In this paper, our concentration has been placed on the point-to-point communication paradigm. We are planing to extend the current model in the context of data broadcast [30], so that items of interest to most mobile clients would be broadcast from a database server while the items of interest to single client would be disseminated over dedicated channels on demand. For example, database indexing information can be broadcast in wireless channels and mobile clients can use this information for efficient evaluation of queries. We would like to further generalize the query processing model for a hybrid broadcast and point-to-point communication environment. This may result in much better utilization of wireless communication links. The design of the cache table might also need to be modified in order to take advantage of the broadcast data.

Acknowledgments

We would like to thank the anonymous referees who have provided valuable feedbacks to improve this paper. The authors of this research are partially supported by the Hong Kong Research Grant Council under grant number HKP72/95E and by the Hong Kong Polytechnic University under grant number 350/570.

References

1. A. Adya, "Transaction management for mobile objects using optimistic concurrency control," Master's thesis, Massachusetts Institute of Technology, July 1994.
2. AT&T Bell Laboratories. Ode 4.1 User Manual.
3. T. Atwood, J. Dubl, G. Ferran, M. Loomis, and D. Wade, The Object Database Standard: ODMG-93, Morgan Kaufmann 1993.
4. B.R. Badrinath and T. Imielinski, "Sleeper and workaholics: Caching strategies in mobiles environments," in Proceedings of the ACM SIGMOD International Conference on Management of Data, ACM 1994, pp. 1–12.
5. P.A. Bernstein, V. Hadzilacos, and N. Goodman, Concurrency Control and Recovery in Database Systems, Addison-Wesley: Reading, MA, 1987.
6. A.F. Cardenas, "Analysis and performace of inverted data base structures," Communications of the ACM, vol. 18, no. 5, pp. 253–258, 1975.
7. M. Carey, M. Franklin, M. Livny, and E. Shekita, "Data caching tradeoffs in client-server DBMS architectures," in Proceedings of the ACM SIGMOD International Conference on Management of Data, 1991, pp. 357–366.
8. M.J. Carey, M.J. Franklin, and M. Zaharioudakis, "Fine-grained sharing in a page server OODBMS," in Proceedings of the ACM SIGMOD International Conference on Management of Data, 1994, pp. 359–370.
9. B.Y.L. Chan, "Cache consistency management in mobile distributed environment," M.Phil. thesis, The Hong Kong Polytechnic Unversity, July 1998.

10. B.Y.L. Chan, A. Si, and H.V. Leong, "Cache management for mobile databases: design and evaluation," in Proceedings of the 14th International Conference on Data Engineering, February 1998, pp. 54–63.

11. J. Chim, R.W.H. Lau, H.V. Leong, and A. Si, "Multi-resolution cache management in digital virtual library," in Proceedings of IEEE Advances in Digital Libraries Conference, April 1998, pp. 66–75.

12. M. Choy, M. Kwan, and H.V. Leong, "On real-time distributed geographical database systems," in Proceedings of the 27th Hawaii International Conference on System Sciences, vol. 4, January 1994, pp. 337–346.

13. S.B. Davidson, H. Garcia-Molina, and D. Skeen, "Consistency in partitioned networks," ACM Computing Surveys, vol. 17, no. 3, pp. 341–370, 1986.

14. P.J. Denning, "The working set model for program behavior," Communications of the ACM vol. 11, no. 5, pp. 323–333, 1968.

15. Digital Roamabout Wireless Products, URL:http://www.digital.com.

16. W. Effelsberg and T. Haerder, "Principles of database buffer management," ACM Transactions on Database Systems, pp. 560–595, December 1984.

17. M. Franklin, M. Carey, and M. Livny, "Global memory management in client-server DBMS architectures," in Proceedings of the 18th International Conference on Very Large Data Bases, 1992, pp. 596–609.

18. J.R. Goodman, "Using cache memory to reduce processor-memory traffic," in The 10th Annual International Symposium on Computer Architecture, IEEE, June 1983, pp. 124–131.

19. C.G. Gray and D.R. Cheriton, "Leases: An efficient fault-tolerant mechanism for distributed file cache consistency," in Proceedings of the 12th ACM Symposium on Operating System Principles, 1989, pp. 202–210.

20. R. Gruber, F. Kaashoek, L. Barbara, and L. Shrira, "Disconnected operation in the thor object-oriented database system," in Proceedings of the IEEE Workshop on Mobile Computing Systems and Applications, IEEE, December 1994, pp. 1–6.

21. Y. Huang, A.P. Sistla, and O. Wolfson, "Data replication for mobile computers," in Proceedings of the ACM SIGMOD International Conference on Management of Data, May 1994, pp. 25–36.

22. L.B. Huston and P. Honeyman, "Disconnected operation for AFS," in Proceedings USENIX Symposium on Mobile & Location-Independent Computing, USENIX, August 1993, pp. 1–10.

23. T. Imielinski and B. Badrinath, "Mobile wireless computing: Challenges in data management," Communications of the ACM, vol. 37. no. 10, pp. 18–28, 1994.

24. J. Jannink, D. Lam, N. Shivakumar, J. Widom, and D.C. Cox, "Data management for user profiles in wireless communications systems," Technical Report, Computer Science & Electrical Engineering Department, Stanford University, 1995.

25. A. Kemper and G. Moerkotte, Object-Oriented Database Management: Applications in Engineering and Computer Science, Prentice-Hall, International edition. 1988.

26. J. Kistler and M. Satyanarayanan, "Disconnected operation in the coda file system," ACM Transaction on Computer Systems, vol. 10, no. 1, pp. 3–25, 1992.

27. N. Krishnakumar and A.J. Bernstein, "Bounded ignorance: A technique for increasing concurrency in a replicated system," ACM Transactions on Database Systems, vol. 19, no. 4, pp. 586–625, 1994.

28. W. Lee and D.L. Lee, "Using signature and caching techniques for information filtering in wireless and mobile environments," Special Issue on Databases and Mobile Computing, Journal on Distributed and Parallel Databases, vol. 4, no. 3, pp. 205–227, 1996.

29. H.V. Leong and A. Si, "Data broadcasting strategies over multiple unreliable wireless channels," in Proceedings of ACM International Conference on Information and Knowledge Management, 1995, pp. 96–104.

30. H.V. Leong and A. Si, "Database caching over the air-storage," The Computer Journal, vol. 40, no. 7, pp. 401–415, 1997.

31. T. Mason and D. Brown, lex & yacc, Nutshell Handbook, O'Reilly & Associates, 1991.

32. C. Min, M. Chen, and N. Roussopoulos, "The implementation and performance evaluation of the ADMS query optimizer: Integrating query result caching and matching," In Proceedings of the 4th International Conference on Extending Database Technology, 1994, pp. 323–336.

33. E. O'Neil, P. O'Neil, and G. Weikum, "The LRU-k page replacement algorithm for database disk buffering," in Proceedings of the ACM SIGMOD International Conference on Management of Data, 1993, pp. 297–306.

34. C. Pu and A. Leff, "Replica control in distributed systems: An asynchronous approach," in Proceedings of the ACM SIGMOD International Conference on Management of Data, May 1991, pp. 377–386.

35. A. Si, H.V. Leong, and S.M.T. Yau, "Maintaining page coherence for dynamic HTML pages," in Proceedings of ACM Symposium on Applied Computing, World Wide Web Applications Track, February 1998, pp. 767–773.
36. M. Spasojevic and M. Satyanarayanan, "An empirical study of a wide-area distributed file system," ACM Transaction on Computer Systems, vol. 14, no. 2, pp. 200–222, 1996.
37. W.R. Stevens, UNIX Network Programming, Prentice Hall Software Series, Prentice-Hall, 1991.
38. M.H. Wong, D. Agrawal, and H.K. Mak, "Bounded inconsistency for type-specific concurrency control," Journal on Distributed and Parallel Databases, vol. 5, no. 1, pp. 31–75, 1997.